NASA SP-2000-4519

Partners in Freedom

Contributions of the Langley Research Center to U.S. Military Aircraft of the 1990's

By
Joseph R. Chambers

Monographs in Aerospace History Number 19

The NASA History Series

National Aeronautics and Space Administration
Office of Policy and Plans
NASA History Division
Washington, DC
2000

Library of Congress Cataloging-in-Publication Data

Chambers, Joseph R.
    Partners in freedom : contributions of the Langley Research Center to U.S. military
aircraft of the 1990's / by Joseph R. Chambers.
        p. cm. -- (NASA history series)
    Includes bibliographical references and index.
    1. Langley Research Center. 2. Aeronautics, Military--Research--United
States--History. 3. Airplaces, Military--United States--Design and construction. I. Title.
II. Series.

UG644.H36 .C43 2000
623.7'46'072073--dc21

                                                                                    00-056072

Acknowledgments

I am sincerely indebted to the dozens of current and retired employees of the NASA Langley Research Center who consented to be interviewed and submitted their personal experiences, recollections, and files from which this documentation of Langley contributions was drawn. The following individuals contributed vital information to this effort:

Irving Abel
William J. Alford, Jr.
Jerry M. Allen
Theodore G. Ayers
Donald D. Baals
E. Ann Bare
Bobby L. Berrier
Ralph P. Bielat
James S. Bowman, Jr.
Francis J. Capone
Huey D. Carden
Stanley R. Cole
Mark A. Croom
Richard G. Culpepper
H. Benson Dexter
Robert V. Doggett
Robert H. Daugherty
Moses G. Farmer

Stuart G. Flechner
Charles H. Fox, Jr.
Charles M. Fremaux
William P. Gilbert
Sue B. Grafton
David E. Hahne
James B. Hallissy
Perry W. Hanson
Roy V. Harris, Jr.
Lowell E. Hasel
William P. Henderson
Charles M. Jackson
Lisa E. Jones
Joseph L. Johnson, Jr.
Donald F. Keller
Richard E. Kuhn
John E. Lamar
Laurence K. Loftin, Jr.
Donald L. Loving

Linwood W. McKinney
James C. Newman, Jr.
Luat T. Nguyen
James C. Patterson, Jr.
John W. Paulson, Jr.
Edward C. Polhamus
Richard J. Re
Wilmer H. Reed, Jr.
Rodney H. Ricketts
Charles L. Ruhlin
Maynard C. Sandford
David S. Shaw
M. Leroy Spearman
Raymond D. Whipple
Richard T. Whitcomb
Thomas J. Yager
E. Carson Yates
Long P. Yip

Special thanks to Bobby Berrier and Patricia A. West, who provided their superb technical editing and proofreading skills to the project. I would like to express my special gratitude to Noel A. Talcott and Jeffrey A. Yetter, who provided the inspiration to undertake this activity, and to A. Gary Price, who provided the mechanism.

Thanks also to Percival J. Tesoro for the cover design, Leanna D. Bullock for assistance with the photographs, Peggy S. Overbey for manuscript preparation services, Cathy W. Everett for bibliography verification and formatting, Christine A. Ryan for printing coordination, and Gail S. Langevin for editing and document production coordination.

Ultimately, however, the greatest thanks go to the thousands of current and retired employees of the NASA Langley Research Center, who provided the personal dedication, expertise, and innovation that enabled the legendary contributions covered in this work.

Joseph R. Chambers
Yorktown, VA
April 20, 2000

Preface

For over 80 years, Langley Research Center has exemplified the cutting edge of world-class aeronautics research for civil and military aircraft. Established in 1917 as the nation's first civil aeronautics research laboratory under the charter of the National Advisory Committee for Aeronautics (NACA), Langley initially existed as a small, highly productive laboratory with emphasis on solving the problems of flight for the military and the civil aviation industry. During World War II (WWII), the Langley Memorial Aeronautical Laboratory directed virtually all of its workforce and facilities to research for military aircraft. Following WWII, a more balanced program of military and civil projects was undertaken. The emergence of the Space Age and the incorporation of the NACA and Langley into the new National Aeronautics and Space Administration (NASA) led to a rapid growth of space related research and the cultural change of the old laboratory into a major research center. Today, Langley research efforts encompass critical areas of both aeronautics and space technology.

Throughout its history, Langley has maintained a close working partnership with the Department of Defense, U.S. industry, universities, and other government agencies to support the defense of the nation with fundamental and applied research. Many of the legendary contributions of Langley to military aircraft technology have been discussed and documented by specialists, the media, and historians. Langley contributions to famous military projects such as the aircraft drag cleanup studies of WWII, the advent of supersonic flight and the X-1, the development and tests of the Century-series fighters, the X-15, and many, many others have been archived in detail.

The objective of this particular undertaking is to document the contributions of Langley Research Center to specific military aircraft that were operational in the 1990's. Virtually all military aircraft that participated in Operation Desert Storm, Kosovo, and other peacekeeping missions of this era have Langley technical contributions to their design, development, and support. In some instances Langley research from one aircraft development program helped to solve a problem in another development program. At the conclusion of some development programs, Langley researchers obtained the research models to conduct additional tests to learn more about previously unknown phenomena. These data also proved useful in later developmental programs. Perhaps the most consistent element in all of the research programs is the length of time for the development and maturation of new research concepts before they are implemented in new aircraft. Many of the military aircraft in the U.S. inventory as of late 1999 were over 20 years old. Langley activities that contributed to the development of some of these aircraft began over 50 years prior.

This publication documents the role—from early concept stages to problem solving for fleet aircraft—that Langley played in the military aircraft fleet of the United States for the 1990's. The declassification of documents and other material has provided an opportunity to record the contributions of Langley personnel and facilities and discuss the impact of these contributions on Department of Defense aircraft programs. This review is intended for the general public with an interest in aircraft development. For more technical information about specific aircraft and programs, please see the publications listed in the bibliography.

Readers familiar with NASA and its research centers will note that the former Lewis Research Center is referred to by its new name, Glenn Research Center. The decision to use the new name was made to avoid confusion for those readers less familiar with the NASA centers and to avoid disruptive explanations in the text for all readers.

Contents

Introduction

*"The making available to agencies directly concerned with national defense
of discoveries that have military value or significance...."*

The foregoing statement is an excerpt from the National Aeronautics and Space Act of 1958, which in part established the formal relationship between NASA and those responsible for the defense of the nation. Since its initial operations over 80 years ago, the Langley Research Center has maintained an appropriate priority within its research activities to ensure the quality, timeliness, and applications of its research in accordance with this congressional act. As a result, Langley and the Department of Defense (DOD) have maintained a strong and productive relationship. This critical teamwork has been nurtured by mutual respect and recognition that the close working arrangements have benefited both parties.

Langley's support of specific military aircraft development programs and Langley's fundamental research have provided many benefits to the DOD. By virtue of its independent agency perspective, Langley's assessments of evolving technology and aircraft systems have provided, and continue to provide, the DOD with unbiased analysis, opinions, data, and extremely valuable recommendations for decisions about aeronautical technology issues in its aircraft programs. Langley's staff has been frequently called upon to participate or represent the DOD in early assessments and selections of competing aircraft designs. In addition, Langley's unique wind tunnels, simulators, and computational facilities have been extensively utilized for evaluations and development of military aircraft. The DOD benefits from the extensive experiences and corporate knowledge of the Langley staff as a result of Langley's participation in a vast number of aircraft development programs. The multidiscipline expertise at Langley is viewed as a unique capability and resource for the nation.

The DOD frequently requests the participation of Langley on review boards and accident investigations. The DOD also recognizes that technical problem-solving exercises during aircraft development programs frequently do not provide the fundamental understanding or design tools necessary to avoid similar problems in future aircraft programs. Therefore, the DOD encourages Langley to conduct follow-on research and frequently assists NASA in obtaining the resources required for these efforts.

Finally, the DOD and its supporting industries value the innovative concepts and technical capabilities provided by the ongoing fundamental research programs at Langley. Breakthrough concepts in aeronautics take years of dedicated research to bring to maturity and readiness for application. Many examples are discussed herein where Langley conceived, assessed, and reduced the risk of application of enabling technology when that technology was critically needed for the development of major DOD aircraft programs.

Providing support to the DOD in the development of military aircraft has provided Langley with very significant benefits. By responding to DOD's requests, Langley maintains relevance in research endeavors and receives test assets, valuable data, and validation of research concepts. For example, DOD normally provides wind-tunnel models for assessments in Langley's unique wind-tunnel facilities. (For more information about selected Langley wind tunnels, see the appendix.) After the critical objectives of aircraft development programs have been met, these models are frequently made available to Langley researchers for generic research and assessments of advanced concepts, which provides a mechanism for research studies that might not be otherwise funded within the Langley budget.

By participating in problem-solving exercises with DOD, Langley researchers are exposed to a myriad of real-world requirements and constraints, thereby gaining valuable awareness of limitations in designing and conducting research within specific technical disciplines. Researchers also experience the implication of off-design flight conditions on new technologies, and thus obtain a much broader outlook for future research opportunities. Another valuable benefit to Langley is the application and validation of research concepts and technical analysis methods. By becoming a team member in DOD aircraft development programs, Langley researchers obtain highly valued flight-test data and feedback that is unaffordable within NASA budget constraints.

The single most important benefit of the DOD and Langley partnership, however, is the application of new technologies to this nation's first-line military aircraft, thereby helping to ensure the continued supremacy of the airpower of the United States.

Langley Contributions to Selected Aircraft

The recent declassification of many documents detailing research conducted at Langley during the 1960's, 1970's, 1980's, and 1990's has permitted a unique opportunity to collate and summarize the contributions of Langley researchers to several types of aircraft.

The following pages document specific contributions of the NASA Langley Research Center to U.S. military aircraft. The discussion includes fighters, transports, trainers, missiles, and remotely piloted vehicles. The information, which has been gathered from documents, personal interviews, and other sources, emphasizes the highlights of Langley's support of specific programs. Research on developing aircraft is extensive and exhaustive. In many cases, additional projects or research was conducted on specific aircraft; however, not all activities are covered herein.

The Langley research culture that produced these significant results flourished because of team participation and the personal dedication and contributions of many, many individuals. Despite the dangers of citing certain individuals and omitting others, names have been included for further guidance to those interested in expanding the information.

The material is presented in two formats. These formats are

- Single-page (back and front) overviews that summarize the major contributions of Langley to a specific aircraft
- Multipage detailed discussions of individual contributions to a specific aircraft

The discussion is intentionally at a level that can be understood by the nontechnical reader; however, additional publications are provided in the bibliography for those seeking more detailed technical information.

BAI Exdrone BQM-147A

SPECIFICATIONS

Manufacturer
    BAI Aerosystems, Inc.
Date in service
    Late 1980's
Number built
    Over 500 to date
Type
    Remotely piloted vehicle
Crew
    Remotely piloted
Engine
    8-hp 2-stroke gasoline
    engine, 2-blade wooden
    propeller

MISSION

Reconnaissance, communications jamming, and delivery of nonlethal payloads

USERS

U.S. Marine Corps and U.S. Naval Air Warfare Center

DIMENSIONS

Wingspan . . . . 8.2 ft
Length . . . . 6.1 ft
Height . . . . 1.7 ft
Wing area . . . . 21.4 sq ft

WEIGHT

Empty . . . . 45 lb
Max payload . . . . 46 lb

PERFORMANCE

Cruise speed . . . . 90 mph
Range . . . . 50 mi
Endurance . . . . 2 hr

HIGHLIGHTS OF RESEARCH BY LANGLEY FOR THE EXDRONE

1. At the request of the U.S. Marine Corps, Langley conducted wind-tunnel and flight tests to solve unacceptable Exdrone low-speed flight characteristics.

2. Langley identified configuration modifications to the wing airfoil, control surfaces, wing leading edge, and vertical tail that dramatically improved low-speed flight characteristics.

The worldwide deployment of remotely piloted vehicles (RPV's) by military forces has dramatically increased. Several factors, such as elimination of threats to human pilots in hostile environments, cost-effectiveness for certain missions, and stealth for some applications, have stimulated the use of RPV's. The field-launched Exdrone RPV has proven to be an extremely effective battlefield option for the U.S. Marine Corps. During Operation Desert Storm, Exdrone was pulled from a research and development program. The Exdrone was pressed into reconnaissance service to find attack routes through Iraqi defenses, thus allowing a rapid ground advance into Kuwait. The NASA Langley Research Center made a critical and timely contribution during the development of the Exdrone that enabled highly successful deployments of the vehicle.

In the mid-1980's, prior to the involvement of Langley, BAI Aerosystems, Inc. manufactured the Exdrone RPV in accordance with a technical specification generated by Johns Hopkins University Applied Physics Laboratory. This version of the Exdrone was for an earlier mission (jamming of communications) that required high dash speeds, therefore, the configuration was not optimized for the low-speed flights that are necessary for extended reconnaissance missions. In fact, at slow speeds the vehicle was extremely difficult to fly and several crashes occurred during field trials simulating the reconnaissance task.

In response to an urgent request from the Marine Corps in 1988, Langley conducted wind-tunnel and flight tests of the Exdrone to determine how the vehicle could be modified to enhance the flight characteristics in low-speed flight. Changes to the wing airfoil, control surfaces, wing leading edge, and vertical tail that were recommended by the Langley staff resulted in a configuration with outstanding low-speed flight characteristics. The changes were incorporated into the BQM-147A Exdrone. The modified vehicle was an immediate success in reconnaissance missions in Operation Desert Storm.

The overall aerodynamic configuration of the Exdrone has not been revised since Langley's study; however, the vehicle payloads and mission applications continue to be upgraded. The latest vehicle is referred to as the Dragon Drone. As technological innovations continue to evolve candidate payloads into smaller, more efficient packages that are suitable for small RPV's, the Dragon Drone's capabilities will likewise continue to be upgraded. The nucleus of the concept, however, remains the modified configuration that owes much to the improvements recommended by Langley's insightful researchers and their problem-solving efforts.

LANGLEY CONTRIBUTIONS TO THE EXDRONE BQM-147A

Background

BAI Aerosystems, Inc. (BAI) of Easton, Maryland initially manufactured the Exdrone air vehicle in the late 1980's in accordance with a technical specification generated by the Johns Hopkins University Applied Physics Laboratory (JHU-APL). JHU-APL's delta-wing configuration was powered by a tractor propeller propulsion system. This configuration was effective for an earlier mission for jamming of communications that required high dash speeds, therefore the configuration was not optimized for the low-speed flights that are necessary for extended reconnaissance missions. During evaluation for potential reconnaissance applications, the U.S. Marine Corps found that the vehicle exhibited poor stability and control characteristics and had a tendency toward severe lateral-directional instability near the stall, which resulted in numerous crashes.

At the request of the Marine Corps in 1988, exploratory wind-tunnel and flight-test investigations of the Exdrone were conducted at the Langley Research Center by flight dynamics specialists under the lead of Joseph L. Johnson, Jr. Lead researcher for the study was Long P. Yip.

Langley Research Efforts

Results of wind-tunnel tests of the baseline Exdrone configuration in the Langley 12-Foot Low-Speed Tunnel identified several aerodynamic deficiencies that contributed to the unacceptable low-speed flight characteristics experienced in the Marine Corps evaluations. For example, longitudinal control was insufficient to trim the Exdrone to the high-lift conditions required for low-speed flight. Yip recommended an increase in the chord of the elevator and thereby provided almost three times as much lift for low-speed flight. To cure a wing-dropping tendency at high lift, Yip recommended a leading-edge "droop" modification to the outer wing panels. Increasing the rudder area and the size of the vertical tail solved deficiencies in directional stability and control. The original configuration displayed a bad combination of lightly damped rolling and yawing motions (sometimes called Dutch roll), but Yip eliminated this motion by increasing directional stability with the increased vertical tail and by adding wingtip skids. The wingtip skids also provided better flow over the ailerons and served as landing skids, which replaced the original drag-producing wire skids.

Langley researchers Long P. Yip and David Fratello with modified Exdrone during flight-test evaluation.

One noteworthy modification contributed by NASA was a sawtooth notch on the wing leading edge near the wingtip. This notch and the leading-edge droop mentioned earlier were an outgrowth of the Langley General Aviation Stall-Spin Program. Yip and his peers had previously recommended the combination to certain general aviation companies for increasing stall departure and spin resistance with a minimal drag penalty. Thus, NASA research to improve light aircraft provided a critical improvement for military applications. As a result of these improvements, the Exdrone is easily flown by inexperienced pilots and very forgiving during training maneuvers.

The results of radio-controlled flight tests conducted by Langley researchers at the Langley Plum Tree Test Site, located in nearby Poquoson, VA, showed that the modified configuration had excellent longitudinal and lateral-directional flight characteristics. The configuration was very maneuverable and responsive to control inputs, exhibited good damping characteristics, and was easily flyable through the stall with no departure tendencies.

The modifications recommended by Langley were endorsed and applied to the Exdrone design, and the overall aerodynamic configuration of the Exdrone vehicle has not changed since these modifications.

Subsequent Exdrone Vehicle Applications

In early 1990, approximately 30 of the newly configured BQM-147A Exdrone reconnaissance RPV's were sent to Operation Desert Storm. The Exdrones were successfully used to map Iraqi minefields and bunkers, which allowed the Allied ground forces to slip through in darkness. During the 1990's, several hundred Exdrone air vehicles were produced by BAI under contract to the U.S. Uninhabited Air Vehicle Joint Project Office (UAV-JPO). These vehicles were sent to Army and Marine Corps units as cost-effective devices for familiarizing new users with the benefits of tactical unmanned reconnaissance systems.

Throughout the 1990's, the UAV-JPO directed BAI, the Department of the Navy, and the Army Research Laboratory to test and incorporate numerous payloads and system upgrades for the basic Exdrone aircraft. Some of those upgrades include Global Positioning System (GPS) (Exdrone was among the first military aircraft certified to use GPS for navigation), a communications relay, the Tactical Remote Sensor Suite (TRSS), an infrared and several other versions of down-looking reconnaissance sensors, and a parachute recovery system to protect the higher value payloads.

In early 1997, the Marine Corps provided funds to BAI to create a new configuration of the Exdrone (referred to as Dragon Drone) for the Hunter-Warrior Advanced Warfighting Experiment at Twenty-Nine Palms, CA. Tests performed at NASA Wallops Flight Facility successfully identified and led to the elimination of onboard vibration interference with reconnaissance payloads. A belly-mounted pan-tilt-zoom television camera produced by BAI was then installed into Dragon Drone with several other improvements.

During the Hunter-Warrior Experiment, Dragon Drones proved highly effective in identifying enemy command locations and troop movements at distances of up to 30 mi. Because of the success of the Hunter-Warrior Experiment, the Marine Corps funded the development of other upgrades to the Dragon Drone system, including

- Belly-mounted television camera equipped with laser range finder for enhanced targeting

- Belly-mount infrared camera equipped with laser range finder for enhanced targeting at night
- Pneumatic launcher and deck-mounted net recovery system to enable shipboard use
- Belly-mounted dispenser for insertion of nonlethal warfare agents (i.e., tear gas)

The Marine Corps' success with the Dragon Drone has sparked worldwide interest in the system. BAI Aerosystems, Inc. received the first international order for a Dragon Drone system from the Bahrain Defense Force (BDF) with delivery in late July 1999. A small Dragon Drone system is being leased to the Australian Marines for evaluation. The Ministry of Defense of the United Kingdom is investigating the feasibility of Dragon Drone providing turnkey fulfillment of their tactical UAV requirements. Several U.S. organizations, including the Air Force and Coast Guard, are actively pursuing these systems.

Clearly, the future for the Dragon Drone is bright. As technological innovations continue to evolve candidate payloads into smaller, more efficient packages that are suitable for RPV's, the Dragon Drone's capabilities will likewise continue to be upgraded. The nucleus of the configuration, however, remains the vehicle configuration that came from the timely and responsive improvements recommended by insightful researchers at Langley.

An Exdrone undergoes weight and balance tests.

Boeing AV-8 Harrier

SPECIFICATIONS

Manufacturer
  Boeing
Subcontractor
  British Aerospace
Date in service
  November 1983
Type
  V/STOL close support
Crew
  One
Engine
  Rolls Royce F402-RR-408

USERS

U.S. Marine Corps, British Navy, British Air Force, Spanish Navy, and Italian Navy

DIMENSIONS

Wingspan . . . . 30.3 ft
Length . . . . 47.75 ft
Height . . . . 11.6 ft
Wing area . . . . 230.0 sq ft

WEIGHT

Empty . . . . 14,867 lb
Max VTOL . . . . 20,595 lb

PERFORMANCE

Max speed . . . . . Mach number
                    of 1.0

HIGHLIGHTS OF RESEARCH BY LANGLEY FOR THE AV-8

1. Free-flight model tests and powered-model studies of the P.1127 by Langley instilled confidence in Hawker's commitment and converted many skeptics in the United Kingdom.

2. Flight tests and coordination by the Langley chief test pilot provided guidance to the British and kept the U.S. and NASA in partnership with leading-edge vertical and short takeoff and landing (V/STOL) developments.

3. Langley flight research and engineering development of vectoring in forward flight (VIFF) provided the Harrier with unprecedented maneuver options in air-to-air combat.

4. Langley provided independent tests and analysis of a deficient competing V/STOL fighter concept (XFV-12A).

5. Langley provided wind-tunnel database for design of AV-8B wing, including beneficial flap and power effects and supercritical airfoil which provide tremendous increase in STOL load carrying capability over previous versions of the AV-8.

The Langley Research Center has played a key role for over 40 years in the highly successful development of the Boeing (formerly McDonnell Douglas) AV-8 Harrier—the most advanced vertical and short takeoff and landing (V/STOL) high-performance aircraft in the world. Langley has been involved in the development of the Harrier since the conceptual and evolutionary stages and continuing through the P.1127 prototype, the Kestrel (XV-6A), the AV-8A, and the current AV-8B used by the U.S. Marine Corps.

Langley's initial involvement with the Harrier began when the British Hawker Company designed a V/STOL prototype known as the P.1127 in 1957. Unfortunately, Hawker's revolutionary design was met with disinterest by the British government and a lack of government funding to proceed into development. At that time, experts at Langley had conducted extensive research on numerous competitive concepts for V/STOL flight, including aircraft-tilting (tail sitters), thrust-tilting (tilt rotors), thrust-deflection (deflected slipstream), and dual-propulsion (lift-cruise engines) concepts. The simplicity and elegance of the rotatable nozzle vectored-thrust concept of the P.1127 so impressed Langley management and researchers that a formal agreement for cooperative testing was initiated with Hawker under the Mutual Weapons Development Program of NATO. Free-flying tests of a 1/6-scale model were conducted in the Langley 30- by 60-Foot (Full-Scale) Tunnel to evaluate flying characteristics and to demonstrate the ease of converting between hovering flight and conventional wing-borne forward flight. These dramatic model flight tests provided confidence to the Hawker test pilots and design team and helped sway the opinions of skeptics in the United Kingdom. Transonic wind-tunnel tests of a powered P.1127 model were also conducted in the Langley 16-Foot Transonic Tunnel. In recognition of his preeminent position as the world leader in flight testing of V/STOL aircraft, Langley chief test pilot John P. (Jack) Reeder was invited to evaluate the P.1127 aircraft and became a strong supporter of the concept.

Following cooperative flight-test evaluations of the second-generation P.1127, the Kestrel, by military of the United Kingdom, the United States, and the Federal Republic of (West) Germany, Langley was provided with two aircraft. These aircraft were used to conduct extensive flight tests to develop more time- and fuel-efficient instrument approach procedures. Arguably, Langley's most significant contribution to the Kestrel was the flight research and engineering development of the vectoring in forward flight (VIFF) concept for enhanced maneuverability in air-to-air combat. This capability provides the Harrier pilot with unique maneuver options. Other Langley contributions included wind-tunnel databases to optimize wing-flap-nozzle aerodynamic interactions and the supercritical wing design methodology for the AV-8B. Langley's leadership role in V/STOL technology was transferred to the NASA Ames Research Center in 1973 where NASA support for the AV-8 series continues. Langley supports the program in technical disciplines that are unique to Langley, such as spin tunnel tests of new variants.

LANGLEY CONTRIBUTIONS TO THE AV-8

Background

In the early 1950's, the Langley Research Center was recognized worldwide as a leader in fundamental and applied research on vertical takeoff and landing aircraft. Leaders in the Langley research efforts included John P. Campbell, Richard E. Kuhn, John P. (Jack) Reeder, and Marion O. McKinney. The challenge of providing efficient vertical flight with minimal penalties and adequate payload produced a myriad of candidate concepts, including aircraft-tilting (tail sitters), thrust-tilting (tilt rotors), thrust-deflection (deflected slipstream), and dual-propulsion (lift-cruise engines) concepts. The Langley researchers had accumulated in-depth experience with each concept and had identified the limitations and complexities that constrained the satisfactory growth of vertical and short takeoff and landing (V/STOL) aircraft. In view of this vast experience and their innovativeness and visionary personalities, the researchers were actively sought for assessments and opinions of emerging V/STOL concepts.

In the United Kingdom, Hawker Aircraft Ltd. was privately funding the development of a new V/STOL tactical strike aircraft known as the P.1127. Initial interest from the British government had been lukewarm, and Hawker aggressively pursued potential funding from the Mutual Weapons Development Program (MWDP) of the North Atlantic Treaty Organization (NATO) for development of the revolutionary P.1127 engine. This engine utilized four swiveling nozzles to redirect the engine thrust for vertical or forward flight. The U.S. members of the MWDP were particularly impressed with the P.1127 concept, and with their outspoken leadership, critical development funds were provided to Bristol Siddeley, the engine manufacturer, in June 1958.

The support for the P.1127 project from the U.S. military (particularly the Marine Corps) and NASA has been a key element in the success of the Harrier, which continues to the present day.

Contributions to the P.1127

As Hawker proceeded in the engineering development of the P.1127 from 1959 to 1960, numerous critical issues arose. These critical issues included the design of the flight control system; whether artificial stabilization was required; the lifting capability of the aircraft in ground effect; and the stability, control, and performance of the P.1127 in conventional flight. Perhaps the most daunting question was whether the aircraft could satisfactorily perform the transition from hovering flight (supported by the vertically directed engine thrust) to conventional wing-borne flight. Many skeptics—particularly in the British government—believed that the transition maneuver would be far too complex for the pilot or that the P.1127 would not maintain adequate lift to permit a safe conversion.

John Stack, then Assistant Director of Langley and an active member of the MWDP, regarded the P.1127 as the most significant advance since the achievement of operational supersonic speeds in fighters. Stack directed the Langley team to provide full support to the emerging P.1127 technology by conducting tests in the unique facilities at Langley. Two model test programs were initiated. One, free-flight tests of a 1/6-scale dynamically scaled powered model in the Langley 30- by 60-Foot (Full-Scale) Tunnel, was used to determine the characteristics of the P.1127 in the transition maneuver. The other program used a large-scale powered model (with simultaneous simulation of inlet and exhaust flows) for force tests in the Langley 16-Foot Transonic Tunnel to determine the complex propulsion and airframe interactions over the operational flight envelope. The 1/6-scale model was also used for tests on the Langley Control Line Facility (a

*Shielded by a protective panel, Langley researchers
Robert O. Schade and Louis P. Tosti hover the 1/6-scale free-flight
model in the airflow return passage of the Full-Scale Tunnel.*

large rotating crane equipped with control lines for testing powered models) to determine characteristics during rapid transitions to and from hovering flight. All tests were slated for completion prior to the initial flights of the prototype aircraft in 1961.

Under the direction of Marion McKinney, the free-flight model tests showed that the P.1127 model behaved extremely well when compared with other V/STOL designs tested by Langley. Transitions to and from forward flight were easily performed, and thrust management was relatively simple. Several problems were identified, however, including the fact that the model lacked sufficient lateral control power for satisfactory behavior during the transition. (The control power of the aircraft was increased as a result of these tests.) A tendency to pitch up due to longitudinal instability at high angles of attack was anticipated based on Hawker wind-tunnel tests and was readily apparent in the model flight tests. (This problem was subsequently cured by adding anhedral or droop to the horizontal-tail surfaces of the P.1127 and subsequent variants.) Despite these shortcomings, the P.1127 was judged to be a superior performer by the Langley researchers.

The free-flight model tests were witnessed by leaders of the Hawker design team, including William Bedford, the P.1127 test pilot slated to make the first conversion flights of the aircraft. While at Langley, Bedford flew Langley's variable stability research helicopter, which had been programmed to simulate the control powers and sensitivities of the P.1127 design. The variable stability features of the helicopter provided valuable information that was used by Hawker in the control system design of the P.1127.

Hovering flight tests of the P.1127 were conducted in the United Kingdom on October 21, 1960, followed by the first conventional flight on March 13, 1961. Finally, on September 12, 1961, transition flights both to and from wing-borne to jet-borne flight were accomplished. The overall results of these flight programs agreed remarkably well with the Langley model tests and the helicopter in-flight simulations of 1960. John Stack witnessed transition flights in gusty conditions a week later and referred to them as the smoothest transition of any of the existing crop of V/STOL machines. Stack subsequently noted that the precursor Langley tests had, in fact, indicated that the P.1127 would have better characteristics than any other concept previously investigated.

Perhaps the most important compliment to the Langley contributions prior to the first flights came from Sir Sydney Camm, the Chief Designer of Hawker (designer of the Hawker Hurricane fighter of WW II), who said that the Langley wind-tunnel tests were the most important tests for the P.1127 project prior to flight.

Despite the success of the P.1127 flight program, the British Royal Air Force did not consider the aircraft as a serious strike aircraft, citing an unacceptably small payload capability and low engine thrust. Aggravating the lack of interest, in March 1961 NATO requested proposals for a new V/STOL close-support fighter with supersonic speed capability. The Hawker design team responded with the P.1154, a configuration with twice the thrust, twice the speed, twice the weight, and twice the performance of the P.1127. While pursuing the P.1154, Hawker continued demonstrations of the subsonic P.1127 and kept the program alive.

Langley chief test pilot Jack Reeder evaluated the P.1127 in 1962.

On June 13, 1962, Langley chief test pilot Jack Reeder became the first foreign pilot to fly the P.1127. Despite the fact that he had never flown in the military service, Reeder was vastly experienced, having flown over 177 different aircraft of which 7 were V/STOL research aircraft. His V/STOL flight experience was unrivaled and Hawker eagerly awaited his opinion of the P.1127. Reeder returned to the U.S. filled with enthusiasm for the P.1127 and he influenced many decision makers regarding the potential of the aircraft for military applications.

The Labour government in the United Kingdom cancelled the P.1154 program and instructed the frustrated Royal Air Force to accept an upgraded version of the subsonic P.1127—the Harrier. But first, the P.1127 was developed into an interim version known as the Kestrel. Nine Kestrel aircraft participated in a unique international collaboration that was designed to assess the practicality of V/STOL operations in the field.

Contributions to the Kestrel

Under the leadership of the MWDP, an agreement was signed in late 1961 by the United States, the Federal Republic of (West) Germany, and the United Kingdom to test an improved P.1127 concept in field conditions in the United Kingdom. Changes made to the P.1127 to upgrade it into the Kestrel included a new engine with increased thrust, a new swept wing with more fuel capacity than the P.1127 wing, a drooped horizontal tail, and improved reaction controls. The flight-test evaluations began in 1965. At the end of 9 months of flight evaluations, the squadron pilots gave glowing reports about the flying qualities of the Kestrel.

Jack Reeder was an active participant in the evaluation program, especially in discussions about required improvements in the Kestrel handling qualities. Following the international flight program, Reeder persuaded officials to provide Langley with two Kestrel aircraft for follow-on V/STOL research. Langley test pilots Lee H. Person, Jr. and Perry L. Deal flew the Kestrels (designated XV-6A) at Langley under the leadership of Reeder.

Contributions of the Kestrel flight tests at Langley had profound impact on the operational usage of this unique vehicle in both the powered-lift regime, as well as in conventional maneuvering flight. Extensive flight evaluations of the efficiency of existing Kestrel instrument approach procedures led by researcher Samuel A. Morello identified new methods to permit safer, more fuel-efficient approaches and landings. However, the most valuable contribution made by the Langley team was the flight research and engineering development that permitted the rotatable nozzles to be deflected in maneuvering flight, thereby providing unprecedented maneuvering for air-to-air combat.

In 1969, the Defense Department requested that Langley review and comment on a report written by Dr. John Attinello, an engineer of the Institute of Defense Analyses, that favorably discussed the potential of using thrust vectoring on P.1127-type aircraft to enhance the maneuverability of fighters in air combat. Although the application of vectoring in forward flight (VIFF) was fundamentally attractive, considerable engineering concern existed over potential control requirements, stability characteristics, and the physical well being of the engine in such maneuvers.

The Langley team designed a flight-test program to develop and evaluate the VIFF concept with the Langley Kestrel. Person and Deal were assigned as project pilots and Richard G. Culpepper was assigned as project engineer. Hawker was initially very skeptical and very concerned over the flight-test objectives, with a special concern expressed over the internal air ducts leading to the reaction control "puffers" at the wingtips and

Lee Person and Jack Reeder with the two Kestrel (XV-6A) aircraft assigned to Langley.

tail of the Kestrel. Operation of the reaction controls at high speeds could result in the internal ducts bursting. There was also concern about handling problems, especially at high angles of attack.

The initial flight trials were conducted in straight and level flight, with Person carefully evaluating internal duct pressures, angle of attack, engine exhaust gas temperatures, and handling qualities. The flight envelope was gradually opened up to 250 knots, then up to 450 knots, until the nozzles could be deflected from the horizontal (cruise) position downward through 90 deg to the breaking stop position. In level flight, the deceleration of the aircraft was extremely high and Person reasoned that the abrupt change in speed could be used to force an enemy pilot to over shoot and become the target. The Kestrel also experienced a nose-up trim change when VIFF was used, but forward stick could be used to maintain attitude. During simulated air combat maneuvers, Person would rotate the nozzles all the way down, roll the wings into the turn and create a very rapid, very high decelerating turn. Other Langley pilots who chased the Kestrel in a Langley T-38 aircraft (including Robert A. Champine, noted X-1 pilot) observed that the Kestrel appeared to "turn a square corner" and added their enthusiasm to that of the test crew.

This initial exploration of VIFF was conducted by Langley from January 1970 to the end of June 1970.

Following the completion of the flight trials of VIFF in 1970, attempts to obtain a more modern Harrier aircraft for follow-on VIFF experimentation met with great disappointment because only six development batch Harriers were available, and all six were heavily involved in development flying in the United Kingdom. Former astronaut Neil A. Armstrong, then serving as Deputy Associate Administrator for Aeronautics at NASA Headquarters, used his influence to obtain a British Harrier and have it modified for VIFF research by NASA. A joint VIFF program between NASA and the Royal

Aircraft Establishment was initiated in 1972, and flying in the United Kingdom continued through 1976. Results obtained in flight evaluations against a variety of high-performance adversary aircraft and analyses of evasive maneuvers provided by VIFF against enemy ground-to-air and air-to-air missiles resulted in overwhelming support for VIFF as a valuable tool for the AV-8 pilot.

As far as the U.S. Marine Corps is concerned, the engineering contribution of NASA was invaluable in developing and proving the VIFF concept (ref. 1).

Other studies of the Kestrel included wind-tunnel tests of a large, powered model in the Langley V/STOL Tunnel (later renamed the Langley 14- by 22-Foot Subsonic Tunnel) by a team led by Richard J. Margason with the objective of establishing wind tunnel to flight correlation. Although the test results were published in a NASA report, the flight correlation effort was not undertaken (ref. 6).

Contributions to the AV-8

The U.S. Marine Corps has unquestionably been the strongest supporter of the Harrier concept. Most aircraft development programs, however, are driven by politics, service rivalries, and many factors other than technology. In the fall of 1972, the U.S. Navy issued a request for proposals of the next generation V/STOL aircraft. Unfortunately, the list of candidates did not include any further development of the Harrier. Instead, the Navy favored the North American Rockwell XFV-12A supersonic fighter design. The XFV-12A used a thrust augmentation scheme that diverted the total exhaust flow of the main engine and ejected it through a venetian blind arrangement in the wings to give vertical-lift capability. The concept was considered by many at Langley to be very risky when compared with the proven Harrier approach, but the Navy was prepared to fully fund the development of the aircraft and close out further development of the Harrier. Two activities subsequently transpired that resulted in Langley contributions to the AV-8 program. First, Langley supported test and analysis of the XFV-12A. Second, Langley contributed airfoil design methods and wind-tunnel databases that played a key role in the development of the second-generation AV-8B.

Naval Air Systems Command requested Langley support for the XFV-12A Program. This support included testing a free-flight model in the Langley Full-Scale Tunnel and a spin model in the Langley 20-Foot Vertical Spin Tunnel and conducting a remarkable hovering test evaluation of an XFV-12A prototype at the Langley Impact Dynamics Research Facility (IDRF). This facility, previously known as the Langley Lunar Landing Facility, had been used to train astronauts for the reduced gravity levels of the moon's environment. Interest in using the IDRF was stimulated by the difficulty of mounting the XFV-12A airframe on a more conventional pedestal mount for the hover test.

The results of the free-flight model tests in 1974 in the Full-Scale Tunnel indicated that the projected thrust augmentation for the XFV-12A was considerably less than expected, and the thrust available for vertical flight was insufficient to permit powered-lift flights. Although the configuration flew well in conventional wing-borne flight, the Langley team expressed grave concern over the deficient V/STOL capability of the free-flight model.

In early 1978, tethered hover tests of the full-scale XFV-12A on the IDRF were carried out by a joint team of NASA, Navy, and Rockwell personnel. Richard G. Culpepper served as the lead Langley engineer for the investigation. The IDRF had undergone major modifications to permit static and dynamic tethered hover tests for powered V/STOL aircraft. During 6 months of tests, it became apparent that major deficiencies existed in the XFV-12A for hovering flight, including marginal vertical thrust. Although

Langley researcher William Newsom with the XFV-12A free-flight model with augmentor doors on wing and canard open.

XFV-12A aircraft mounted for tethered hover flights on Langley Impact Dynamics Research Facility.

the augmentation of flow at the wing augmentors was as predicted, large losses in the internal ducting and corners of the propulsion system seriously degraded the net thrust to the extent that only 75 percent of the weight of the vehicle could be supported in attempts to hover. The results of the tests at Langley influenced the Navy's decision to cancel the XFV-12A Program.

Meanwhile, faced with the potential end of the Harrier program, McDonnell Douglas and its partners launched a major redesign effort to provide a significant improvement in the V/STOL capability of the AV-8A. McDonnell Douglas engineers drew on two fundamental research efforts at Langley to assist them in redesigning the AV-8A into the AV-8B. Under the leadership of Richard E. Kuhn, Langley researchers in the Langley V/STOL Tunnel conducted systematic wind-tunnel studies of the aerodynamic interactions that occur between rotatable fuselage-mounted nozzles and a high wing with a trailing-edge flap. The test variables included a range of geometric relationships between these components and showed that the resulting total lift for an aircraft similar to the Harrier could be significantly impacted (both favorably and unfavorably) by the positioning of these elements. Drawing on this database, the McDonnell Douglas engineers arrived at the current AV-8B wing-nozzle-flap configuration, which resulted in an increase of more than 6,000 lb of lift beyond that produced by the AV-8A arrangement. In addition, McDonnell Douglas used methods that had matured from the research of Dr. Richard T. Whitcomb for supercritical airfoils. The resulting AV-8B wing design has a thicker wing with better performance at high speeds, better fuel consumption, and provides an increase in internal fuel capacity of over 40 percent.

After the decision was made by NASA Headquarters in 1973 to consolidate all V/STOL research under the leadership of the NASA Ames Research Center, additional wind-tunnel and V/STOL flight research on the Harrier was conducted by Ames, and close NASA involvement in the AV-8 Program continues today.

Langley continues to support the program in areas unique to Langley expertise and facilities. For example, spin tunnel tests were conducted at Langley for the AV-8B in 1984, as the external configuration, armament, and other important factors changed in the AV-8 fleet.

Boeing C-17 Globemaster III

SPECIFICATIONS

Manufacturer
  Boeing
Date in service
  January 1995
Type
  Transport
Crew
  Three
Engine
  Pratt & Whitney F117-PW
  100 turbofan

USER

U.S. Air Force

DIMENSIONS

Wingspan . . . . 169.8 ft
Length . . . . 174.0 ft
Height . . . . 55.1 ft
Wing area . . . . 3,800 sq ft

WEIGHT

Empty . . . . 277,000 lb
Max take-off . . . . . . 585,000 lb

PERFORMANCE

Max speed . . . . . Mach number
                              of 0.77
Range . . . . 4,741 n mi

HIGHLIGHTS OF RESEARCH BY LANGLEY FOR THE C-17

1. Langley conceived, researched, and developed an externally blown flap concept that permits the C-17 to make slow, steep approaches with heavy payloads.

2. The C-17 uses supercritical wing technology, developed at Langley, which enhances range, cruising speed, and fuel efficiency at transonic cruise conditions.

3. The C-17 employs winglets, conceived and developed at Langley, for better cruise efficiency with a reduced wing span.

4. Fly-by-wire technology, used by the C-17 as a lighter weight replacement for a hydraulic control system, was initially researched at Langley.

5. Research led by Langley on the development of composite materials enabled the C-17 to employ advanced composites for significant weight savings.

6. Langley conducted flutter clearance tests for the C-17 wing and winglet configuration.

7. Langley conducted fundamental research on deep-stall characteristics of T-tail aircraft, providing McDonnell Douglas with background for the development of the C-17 angle-of-attack limiting system.

The C-17 is the newest airlift aircraft to enter the Air Force inventory. The C-17 is capable of rapid strategic delivery of troops and all types of cargo to main operating bases or directly to forward bases in the deployment area. The aircraft is also able to perform theater airlift missions when required. Using advanced aerodynamics and an innovative NASA powered-lift concept, the C-17 combines the load carrying capacity of the C-5 with the short takeoff and landing performance of the C-130. McDonnell Douglas was recognized in 1994 for the innovative design of the C-17 with the prestigious Collier Trophy, which is awarded annually for the greatest achievement in aviation in the United States.

Fundamental and applied aeronautics research conducted at Langley in the areas of advanced high-lift systems, aerodynamics, advanced composites, and aeroelasticity contributed to the success of the C-17. Decades of Langley research efforts had conceived, developed, and matured emerging concepts and design guidelines that helped McDonnell Douglas produce this outstanding military transport. Langley facilities associated with research for the C-17 included the 30- by 60-Foot (Full-Scale) Tunnel, the 300-MPH 7- by 10-Foot Tunnel, the 16-Foot Transonic Dynamics Tunnel, the 8-Foot Transonic Pressure Tunnel, piloted simulators, and the Structures and Materials Laboratory.

In an acknowledgement of Langley's involvement in the C-17, McDonnell Douglas and the Air Force brought a C-17 to Langley on May 24, 1996, for a special ceremony to thank the employees for their contributions to the design of the C-17.

LANGLEY CONTRIBUTIONS TO THE C-17

The CX Competition

In the late 1970's, the U.S. military recognized a growing demand for rapid deployment of military forces and equipment that would exceed the capabilities of the existing C-141, C-5, and C-130 fleets. Early in 1980, the Department of Defense (DOD) issued a request for proposals (RFP) for a new Cargo Experimental (CX) Program. Boeing, Lockheed, and McDonnell Douglas submitted variants of civil transports, derivatives of the prototype YC-14 and YC-15 aircraft, and completely new aircraft in response to the RFP. In August 1981, the Air Force announced that it had selected the Douglas Aircraft Company Division of McDonnell Douglas to develop the CX, now known as the C-17.

Langley's contributions to the development of the C-17 included years of consultation and cooperative research with the Douglas team, providing unique test facilities, and several innovative technological concepts.

The Externally Blown Flap Concept

The specifications for the C-17 transport required advanced concepts for superior short takeoff and landing (STOL) performance. As a result of Langley research on powered high-lift systems for over 35 years, the innovative externally blown flap (EBF) concept had matured to the point that it could be incorporated in the C-17 transport. The EBF enables the C-17 to make slow, steep approaches with heavy cargo loads to touch down precisely on the spot desired on limited runway surfaces. Because of this technology that was developed by Langley, the C-17 can carry the same loads as a C-5 and use the same airfields as a C-130.

John P. Campbell of Langley conceived the EBF concept in the mid-1950's as a relatively simple approach to augment wing lift for low-speed operations. In this concept, the exhaust from pod-mounted engines impinges directly on conventional slotted flaps and is deflected downward to augment the wing lift. The magnitude of lift augmentation is extremely large, and the resulting lift can be as much as twice the value for a conventional aircraft. However, no serious consideration was given to the EBF concept initially because of the severe high-temperature impingement on the wing and flap surfaces from the turbojet (no bypass or fan flow) engines used at that time. Also, the relatively small mass flow from such engines was a limiting factor for lift augmentation. In addition, considerable concern was expressed over potential control problems in the event that an engine became inoperative during flight at low speeds with high-power settings. With the advent of turbofan engines, however, the efflux from the engines was relatively cool, and large quantities of air became available for increased airflow through the flaps. The turbofan engine, therefore, provided the breakthrough mechanism that permitted Langley researchers to evolve and mature the applications of the EBF concept.

At Langley, extensive wind-tunnel tests in the Langley 30- by 60-Foot (Full-Scale) Tunnel and the Langley 300-MPH 7- by 10-Foot Tunnel explored the fundamental lift-augmentation capability of the EBF for various wing-engine nacelle-flap combinations. These tests defined the optimum wing and flap geometries for application to aircraft configurations. Led by Joseph L. Johnson, Jr., a team of researchers in the Full-Scale Tunnel studied the aerodynamic performance of two- and four-engine EBF transport configurations. The scope of their research included detailed studies of projected aircraft performance, potential control problems and solutions, and design guidelines for the general geometric layout of the aircraft. The advantage of using a T-tail empennage con-figuration was identified as a desirable approach to avoid the large local downwash angles experienced at conventional tail locations due to the increased wing circulation

Langley researcher John P. Campbell (l) and Gerald G. Kayten (r) of NASA Headquarters inspect a free-flight model of an EBF configuration in the Langley Full-Scale Tunnel.

produced by the EBF concept. The engine-out control issue was addressed and solved by defining the vertical tail and rudder size required to maintain directional control.

Dynamically scaled models of transport configurations vividly demonstrated the high-lift potential of the EBF concept, while exhibiting satisfactory stability and control characteristics, in free-flight wind-tunnel tests. Aerodynamic data generated by these studies were then used for piloted-simulator studies of EBF transports at Langley, during which more realistic evaluations of the handling qualities could be conducted and flight control systems could be configured. Throughout the years that the EBF concept developed and matured, Langley researchers coordinated with U.S. industry and DOD. Langley provided briefings and invaluable technical reports that summarized the findings of the research studies and the revolutionary capability offered by this new concept in aeronautics.

In the late 1960's, the progress on the EBF concept had advanced considerably, and most of the major issues had been adequately addressed and resolved. The concept was now ready for flight evaluation on an actual aircraft. Langley advocated for a special flight demonstrator program and conducted several studies of the feasibility of modifying existing aircraft as appropriate test beds. The Douglas A3 Skywarrior twin-engine, high-wing aircraft operated by the U.S. Navy appeared to be a desirable option; however, a new DOD program known as the Advanced Tactical Transport Program superseded the Langley plans.

The YC-15

In 1972, the Air Force issued a request for proposals (RFP) for an Advanced Medium STOL Transport (AMST) that would ultimately replace the C-130 tactical transport. The program emphasized innovative technologies and the capability of conducting STOL operations from 2,000-ft runways. Boeing and McDonnell Douglas were each awarded preliminary design contracts for the construction and testing of two transport prototypes, respectively designated YC-14 and YC-15.

Boeing based their YC-14 design on another powered-lift concept known as upper surface blowing (USB), which had also been developed at Langley in efforts led by Joseph Johnson and Oran W. Nicks, Deputy Director of Langley. In the USB concept, the engines were mounted so that the exhaust spread over the upper surface of the wing for enhanced circulation and lift augmentation in STOL operations.

McDonnell Douglas used the EBF concept with a four-engine configuration and large double-slotted flaps that extended over 75 percent of the total span for the YC-15 prototypes. The first of two YC-15 prototypes made its first flight on August 26, 1975, and was joined by the second prototype in December of that year. The YC-15 demonstrated exceptional STOL performance in its flight-test program with an approach speed of only 98 mph and a field length of 2,000 ft at a landing weight of 150,000 lb.

During the flight-test program for the YC-15, Langley researchers participated in evaluations and analysis of STOL capabilities, including an assessment of lift augmentation in ground effect. In addition to participating in the flight tests of the YC-15, Langley conducted a cooperative wind-tunnel test in the Langley 8-Foot Transonic Pressure Tunnel to evaluate the effectiveness of the emerging winglet concept, which was developed by Langley, on the YC-15 configuration. Dr. Richard T. Whitcomb's winglet designs are small, wing-like vertical surfaces located at each wingtip that enable an aircraft to fly with greater efficiency. Winglets are strategically located at the wingtip to produce a forward force on the aircraft, similar in many respects to the sail on a sailboat.

Although the YC-15 was never placed into production, the experiences gained by McDonnell Douglas in advanced wing design (supercritical airfoil and winglets), the EBF concept, and advanced controls for STOL operations gave the company confidence in future applications of these technologies to the C-17.

YC-15 model in the Langley 8-Foot Transonic Pressure Tunnel for an evaluation of winglets.

One of the McDonnell Douglas YC-15 prototypes in flight.

*The C-17 Supercritical Wing,
Winglets, and Aerodynamic
Studies*

The YC-15 was the first military transport to use supercritical wings, a major innovative technology conceived and developed through wind-tunnel research by Richard Whitcomb at Langley. Whitcomb's supercritical wings incorporate advanced airfoils that enhance the range, cruising speed, and fuel efficiency of aircraft by producing weaker upper-surface shock waves, thereby creating less drag and permitting higher efficiency. McDonnell Douglas subsequently incorporated supercritical wing technology in the C-17 design.

Whitcomb's brilliant development of the winglet concept was another product of research at Langley that was ideally suited for the C-17. The C-17 is a large aircraft with a relatively small wing. The wingspan of the C-17 was dictated by an Air Force requirement for three aircraft to maneuver on a ramp measuring 90 m by 122 m that is connected by a 15-m-wide taxiway. The aerodynamic contribution of the winglets permits the C-17 to employ a shorter wing span while retaining the efficiency of a larger wing span. The C-17 winglets also employ supercritical airfoil sections.

The successful application of winglets to the C-17 also required consideration and analysis of the flutter characteristics of the wing-winglet combination. This area was of particular concern because flow separation in the wing-winglet juncture could provide an aggravating mechanism that might lower the flutter speed to within the flight envelope of the C-17. Led by Charles L. Ruhlin, a Langley, McDonnell Douglas, and Air Force team conducted flutter tests on the C-17 wing-winglet configuration in the Langley 16-Foot Transonic Dynamics Tunnel and successfully cleared the aircraft for flight tests.

Several cooperative wind-tunnel test studies were conducted by McDonnell Douglas and Langley in the National Transonic Facility (NTF) at the Langley Research Center to assess and optimize the cruise aerodynamic performance of the C-17. The unique capability of the NTF to more properly simulate full-scale aerodynamic flight conditions has been an extremely valuable contribution to the program.

*Semispan model for the C-17 wing-winglet configuration in the
Langley 16-Foot Transonic Dynamics Tunnel for flutter tests.*

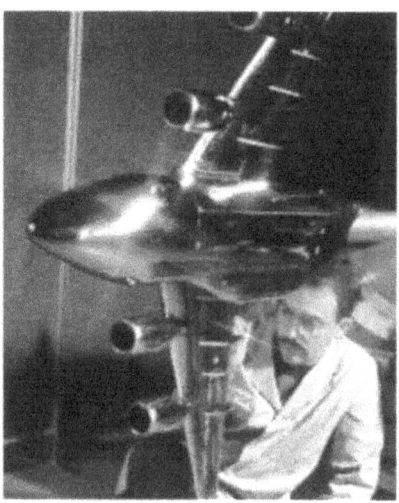

*C-17 model mounted in the National Transonic
Facility at Langley for aerodynamic studies.*

Composite Materials

The initiation of the Aircraft Energy Efficiency (ACEE) Program by NASA in 1976 in response to the energy crisis accelerated the development of concepts to improve the efficiency of advanced aircraft. New methods of producing unique, lightweight materials were one of the major thrusts of the ACEE Program, with emphasis on durable composite materials for aircraft structures. The Structures Division at Langley played a key role in developing the technology, which was ultimately incorporated on several components of the C-17. One of the most valuable Langley contributions in the ACEE Program was the development of graphite-epoxy upper aft rudders for the DC-10. The rudders have accumulated over 500,000 flight hours since they were introduced into regular airline service in 1976, thereby providing extensive experience for applications to other aircraft, including the C-17.

Several major components of the C-17 are made of advanced composites (ailerons, rudders, elevators, vertical and horizontal stabilizers, flap hinge fairings, main landing gear pod panels, and winglets). Over 16,000 lb of composites are incorporated in the design, with composites accounting for about 8 percent of the structure.

Fly-by-Wire Control System

The extensive database and literature produced by pioneering research at Langley on fly-by-wire flight control systems and handling qualities of STOL transports provided McDonnell Douglas with a rich source of information in the development of the C-17 flight control system. Although not directly involved in this facet of the C-17 development program, Langley contributed fundamental research information that helped build the confidence and risk reduction required for the application of the sophisticated C-17 systems.

Avoiding the Deep Stall

The T-tail empennage configuration offers significant aerodynamic advantages over conventional designs. The relatively high location of the horizontal tail places the tail in a relatively undisturbed airflow at normal cruise conditions, thereby maximizing the contribution of the tail to stability and control. Many military and civil design teams have adopted this tail configuration very successfully. However, the application of the T-tail requires consideration of critical aerodynamic factors—especially an analysis in wind-tunnel tests to ensure satisfactory handling characteristics at extreme pitch attitudes. At high angles of attack associated with wing stall, the low-energy wake of the stalled wing can impinge on the horizontal tail and result in a loss of longitudinal stability (pitch up) and markedly reduced longitudinal control effectiveness. As a result of these flow phenomena, the angle of attack can increase to a deep-stall condition, in which the aircraft enters a stable but uncontrollable trim point with a very high rate of descent.

In the early 1960's, a British BAC 111 T-tail transport experienced a fatal accident in which the aircraft entered a deep stall and descended in an uncontrollable condition at a very high rate of descent with an almost horizontal fuselage attitude until impact. Worldwide interest in the causes of this accident resulted in a research program at the Langley Research Center on the behavior of T-tail configurations at high angles of attack. Under the leadership of Robert T. Taylor and Martin T. Moul, extensive wind-tunnel and piloted-simulator studies were conducted to determine the aerodynamic characteristics associated with deep-stall trim conditions and to develop design methods and pilot techniques to recover from such conditions. A large number of aircraft configurations were studied in the Langley 7- by 10-Foot High-Speed Tunnel, with an emphasis on designs with aft-fuselage-mounted engine nacelles. Studies at Langley on the

T-tail deep-stall phenomenon provided a valuable database that has been used extensively in the design of numerous T-tail civil and military transport configurations. Perhaps the most valuable contribution of this work is the guidance and approach it suggests to the designer during early wind-tunnel tests and configuration layout.

The C-17 design team was very familiar with the existing NASA database and design procedures for avoiding the deep-stall problem. The quadruple redundant digital fly-by-wire flight control system of the C-17 provides automatic limiting of angle of attack for high-angle-of-attack conditions, thereby preventing any tendency of the aircraft to enter an uncontrollable deep-stall condition.

Recognition Visit

On May 24, 1996, McDonnell Douglas and the Air Force brought a C-17 of the Air Mobility Command from Charleston Air Force Base, S.C., to Langley as a gesture of thanks to Langley and its employees for their contributions to the design of the new military jet transport. The visit and formal ceremony acknowledged the contributions of all four NASA Aeronautics Centers (Langley, Ames, Dryden, and Glenn) to the design and development of the C-17. The NASA research contributions cited by the visitors included the areas of powered-lift systems, short takeoff and landing control systems, head-up display technology, supercritical wing and winglet, fly-by-wire systems, engine technologies, and composite materials.

C-17 in front of the hangar during a visit at Langley on May 24, 1996.

Boeing F/A-18 Hornet

SPECIFICATIONS

Manufacturer
  Boeing and Northrop Grumman
Date in service
  1983
Type
  Multirole fighter-attack aircraft
Crew
  One or two
Engine
  F-18C . . . . . . .General Electric
  F404-GE-402
  F-18E . . . . . . .General Electric
  F414-GE-400
USERS
U.S. Navy, U.S. Marine Corps,
Canada, Australia, Spain, Kuwait,
Switzerland, Finland, and
Malaysia
DIMENSIONS
Wingspan
  F-18C . . . . 40.4 ft
  F-18E . . . . 44.9 ft
Length
  F-18C . . . . 56.0 ft
  F-18E . . . . 60.3 ft
Height
  F-18C . . . . 15.3 ft
  F-18E . . . . 16.0 ft
Wing area
  F-18C . . . . 400.0 sq ft
  F-18E . . . . 500.0 sq ft
WEIGHT
  Empty
  F-18C . . . . 23,407 lb
  F-18E . . . . 30,500 lb
  Gross
  F-18C . . . . 51,900 lb
  F-18E . . . . 66,000 lb
PERFORMANCE
Max speed . . . . .above Mach
                        number of 1.8

HIGHLIGHTS OF RESEARCH BY LANGLEY FOR THE F/A-18

1. At the request of industry and the military, Langley participated in reviews and assessments from the YF-17 prototype program to the development of F/A-18E/F.

2. Langley provided wind-tunnel research on wing leading- and trailing-edge flap technology for transonic maneuver optimization.

3. Langley contributed to cooperative studies of vortex lift with McDonnell Douglas and Northrop and participated in the selection of the wing leading-edge extension (LEX) shapes for all variants of the F/A-18.

4. Langley provided a solution (porous wing doors) to a severe wing-drop tendency for the F/A-18E/F, thereby helping to avoid potential termination of the aircraft program.

5. Langley helped solve numerous F/A-18 developmental challenges including
    * Cruise performance
    * Flutter clearance
    * High-angle-of-attack stability and control
    * Recovery from unusual out-of-control conditions

6. Langley led the NASA High-Angle-of-Attack Technology Program, which provided valuable information on F/A-18 characteristics and design methodology for future fighters.

The Langley Research Center has been an active participant in the development of the F/A-18 series of aircraft for over 25 years, including contributions to the YF-17, F/A-18A, and the F/A-18E/F. With the exception of the F-111, no other fighter aircraft has been the subject of as many Langley wind-tunnel studies and analytical investigations. Working closely with Boeing (formerly McDonnell Douglas) and Northrop Grumman (formerly Northrop), Langley has contributed to aerodynamic performance, computational fluid dynamics, high-angle-of-attack stability and control, and aeroelasticity. Langley facilities used in the development of the F-18 included the 30- by 60-Foot (Full-Scale) Tunnel, the 12-Foot Low-Speed Tunnel, the 20-Foot Vertical Spin Tunnel, the 8-Foot Transonic Pressure Tunnel, the 16-Foot Transonic Tunnel, the Jet Exit Test Facility, the 16-Foot Transonic Dynamics Tunnel, the 7- by 10-Foot High-Speed Tunnel, the 14- by 22-Foot Subsonic Tunnel, the Differential Maneuvering Simulator, radio-controlled drop models, and computational facilities.

Research contributions by Langley for the F/A-18 have increased lift for aggressive maneuvering, improved cruise performance, insured a high degree of spin resistance, provided flutter clearance, and enabled prompt recovery from out-of-control conditions. Recently, one of the most significant Langley contributions was providing a solution to an uncommanded wing-drop characteristic exhibited by the F/A-18E/F. Boeing was awarded the Collier Trophy in 1999 in recognition of the accomplishments in the F/A-18E/F program.

Langley provided the lead advocacy and technical leadership for a highly successful NASA High-Angle-Of-Attack Technology Program based on the F/A-18 configuration. The program aggressively accelerated progress in design methods for aerodynamics, flight controls, thrust vectoring, alleviation of empennage buffet, and test techniques. Extensive documentation and analysis of F/A-18 characteristics were significant by-products of the program.

LANGLEY CONTRIBUTIONS TO THE F/A-18

Vortex Lift and Maneuvering Flaps

In the early 1960's, the Northrop Company noticed an improvement in the maximum lift of the F-5 aircraft because of a small flap actuator fairing that extended the wing root leading edge. This phenomenon spurred interest in the effects of inboard vortex flows and led to a cooperative NASA and Northrop study, which was conducted in the Langley 7- by 10-Foot High-Speed Tunnel with a group led by Edward C. Polhamus. The cooperative study of hybrid wings centered on the use of relatively large, highly swept wing extensions at the wing-fuselage intersection, which promoted strong beneficial vortex-flow effects. The scope of the study included parametric studies to maximize the lift- and stability-enhancing effects of the wing extension concept, which became known at Northrop as the leading-edge extension (LEX). Studies were also directed at cambering the leading edge of the LEX to suppress the vortex at low angles of attack, and thereby minimize drag at cruise conditions. Northrop applied a large highly swept LEX to the YF-17 prototype aircraft to enhance lift and stabilize the flow over the YF-17 main wing at high angles of attack.

From these initial cooperative studies with Northrop, Polhamus and his associates put together a world-class vortex-lift research program that became internationally recognized for its experimental database, analytical procedures, and aircraft applications. In addition to Polhamus, key members of this team included Linwood W. McKinney, Edward J. Ray, William P. Henderson, John E. Lamar, and James M. Luckring. Their extraordinary research into the fundamentals and applications of vortex flows placed the Langley Research Center in an excellent position to aid the U.S. industry in the design of highly maneuverable advanced fighters. This experienced pool of experts would subsequently provide invaluable guidance and analysis to industry design teams in the development of the F-16 for the Air Force and the F/A-18 for the Navy.

F/A-18 with wing leading-edge flaps deflected and LEX vortices made visible by condensation.

In Vietnam, the lack of maneuverability of U.S. fighters at transonic speeds provided key advantages to nimble enemy fighters. Industry, the Department of Defense (DOD), and NASA were all stimulated to sponsor research to achieve unprecedented transonic maneuverability while maintaining excellent handling qualities. Langley researchers, under the leadership of Polhamus, conducted studies in the 7- by 10-Foot High-Speed Tunnel to obtain near optimum aerodynamic maneuver performance for wings, including the use of fixed and variable camber concepts. Some of the earliest systematic wind-tunnel tests were conducted by the group to determine the most effective geometries for wing leading- and trailing-edge flaps. In addition to tests of aerodynamic performance and stability and control, buffet studies were conducted to understand and develop methodologies for the prediction and minimization of undesirable buffet characteristics. The program was closely coordinated with flight tests of actual high-performance fighters at the NASA Dryden Flight Research Center. Flight evaluations of the effects of maneuver flaps on a YF-17 were also later conducted in the Dryden program.

Numerous discussions with Polhamus and his staff provided valuable guidance to the Northrop design team and the McDonnell Douglas team for the subsequent F/A-18 design. The insight and understanding provided by the broad database from Langley tests permitted development of the extremely effective leading- and trailing-edge flaps used by the YF-17 and the F/A-18. The F/A-18 and similar high-performance fighters use specific, computer-controlled schedules of flap deflection with Mach number and angle of attack for superior maneuverability throughout the flight envelope.

Development of the YF-17

On January 6, 1972, the Air Force issued a request for proposals (RFP) for a Lightweight Fighter (LWF) Program. In March 1972, the Langley Research Center was requested by DOD to participate in assessments and supporting tests of the competing YF-16 and YF-17 designs for the LWF. Langley researchers became members of DOD source evaluation teams to assess and check technical claims by each of the contractors. The sponsoring LWF Program Office requested that certain services of Langley be made available on an equal basis to the two competing teams. This remarkable arrangement provided each team with analysis and support if they desired.

Northrop placed a high priority on superior high-angle-of-attack characteristics and a high degree of inherent spin resistance for the YF-17. The company had also placed priorities in these areas during the development of the F-5 and T-38 aircraft, which had become known for outstanding resistance to inadvertent spins. Langley support was therefore requested for tests in the 30- by 60-Foot (Full-Scale) Tunnel and the 20-Foot Vertical Spin Tunnel.

To provide superior handling qualities at high angles of attack for fighter aircraft, Northrop provided the airframe with the required levels of aerodynamic stability and control characteristics without artificially limiting the flight envelope with the flight control system. This approach proved to be highly successful for the YF-17 and has been adopted by McDonnell Douglas (now Boeing) and used in all variants of the F/A-18 aircraft.

Researcher Sue B. Grafton conducted exhaustive tests in the Full-Scale Tunnel of the YF-17 configuration at high angles of attack in 1973. The results of the Langley tests revealed that Northrop had done an outstanding job in configuring the YF-17 design. The integration of the large LEX surfaces and the placement of the twin vertical tails provided exceptional tail effectiveness at high angles of attack. The small strakes added to the forward fuselage nose by Northrop resulted in extremely high directional stability at high angles of attack. Free-flight tests of the YF-17 model in the Full-Scale Tunnel

Project engineer Sue Grafton with the free-flight model of the YF-17 in 1973.

confirmed the excellent flying characteristics predicted by the wind-tunnel data and provided Northrop with highly positive predictions for upcoming flight tests of the two YF-17 prototypes at Edwards Air Force Base. The wind-tunnel data also formed the basis for piloted simulator studies at Langley and Northrop that helped Northrop design the flight control system for critical high-angle-of-attack conditions.

Spin and recovery tests in the Spin Tunnel also provided positive results for the YF-17. The scope of the tests in the Spin Tunnel was relatively broad for the fast-paced LWF Program and included the determination of the size of the emergency spin recovery parachute that would be required for flight tests. The results of the Spin Tunnel tests showed that the YF-17 would have remarkably good spin and recovery characteristics. In fact, the YF-17's characteristics were the best noted for any fighter configuration to that time.

Although not specifically requested by the Air Force, Langley had already conducted Air Force approved studies of the high-angle-of-attack characteristics of the YF-16 in the Langley Differential Maneuvering Simulator (DMS) and approval was given to conduct similar studies of the YF-17. Under the leadership of Langley researchers Luat T. Nguyen and William P. Gilbert, extensive studies were conducted in the DMS to verify the impressive behavior predicted by the wind-tunnel and free-flight model tests for more realistic air combat conditions. In addition, the simulation was used to refine certain elements of the flight control system for high-angle-of-attack conditions. The results of the simulator investigation showed the YF-17 to be highly maneuverable and departure resistant throughout the operational angle-of-attack range and beyond maximum lift.

A YF-17 prototype in flight with the open
slots in the LEX adjacent to the fuselage.

YF-17 model in the Langley 8-Foot Transonic
Pressure Tunnel for drag assessment studies.

The Langley predictions for the YF-17 behavior were subsequently confirmed in 1974 when two YF-17 prototypes began flight evaluations at Edwards Air Force Base. Handling characteristics at high angles of attack were excellent. The YF-17 could achieve angles of attack of up to 34 deg in level flight and 63 deg could be reached in a zoom climb. The aircraft remained controllable at indicated airspeeds down to 20 knots. Northrop consequently claimed that their lightweight fighter contender had no angle-of-attack limitations, no control limitations, and no departure tendencies within the flight envelope used for the evaluation.

Langley researchers conducted a cruise-drag test of the YF-17 in the Langley 8-Foot Transonic Pressure Tunnel in 1973. The test was initiated when Northrop questioned the high transonic drag levels predicted by the Air Force, which were based on results from other wind tunnels. The Air Force agreed that an independent NASA analysis would be appropriate and requested the test.

Langley researchers found that their results agreed with the Air Force predictions. Researchers then identified the major contributors to drag and provided recommendations to reduce it. Dr. Richard T. Whitcomb reviewed the YF-17 drag results and concluded that the wing design was the major factor in the unexpected drag levels. Whitcomb's suggestion for a wing redesign to solve the problem was unacceptable to Northrop.

F/A-18A TO F/A-18D

Development of the F/A-18

In April 1974, the LWF Program changed from a technology demonstration program to a competition for an Air Force Air Combat Fighter (ACF), and the flight-test programs for the YF-16 and YF-17 were rushed through in a few months instead of the planned 2 years. On January 13, 1975, the Air Force announced that the General Dynamics YF-16 would be the new ACF. Congress decreed that the Navy adopt a derivative of one of the LWF designs as the new Naval Air Combat Fighter to complement the F-14. On May 2, 1975, the Navy announced that the YF-17 design would form the basis of their new F/A-18 fighter-attack aircraft. Northrop, inexperienced in the design of naval fighters, teamed with McDonnell Douglas, which had extensive experience with a highly successful line of naval aircraft, including the F-4 Phantom. McDonnell Douglas became the prime contractor for the F/A-18 with Northrop the prime subcontractor.

The formidable task of converting the land-based YF-17 lightweight day fighter into an all-weather fighter-attack aircraft capable of carrier operations with heavy ordnance loads required significant changes from the earlier configuration. Structural strengthening and a new landing gear design were required for catapult launches and arrested landings. The aircraft gross weight rapidly grew from 23,000 lb for the YF-17 to a projected weight over 33,000 lb.

The required approach speeds for carrier landings resulted in modifications to the wing and LEX surfaces of the YF-17 configuration to provide more lift. McDonnell Douglas consulted the Langley research staff, and several individuals participated in the analysis of wind-tunnel tests that had been conducted at NASA Ames Research Center and McDonnell Douglas facilities. As a result of the analysis, changes were made to the aircraft configuration. The geometric shape of the YF-17 LEX was extended farther forward on the fuselage and the plan view of the LEX was modified to produce additional lift while retaining the good high-angle-of-attack characteristics exhibited by the YF-17.

The deflections of the wing leading- and trailing-edge flaps were increased and the ailerons were programmed to droop in low-speed flight to augment lift. Finally, a "snag" or discontinuity was added to the leading edges of both the wing and horizontal tails to provide more lift.

Formal Navy requests for specific NASA studies in support of the evolving F/A-18 configuration were received and accepted by Langley. The excellent correlation of Langley predictions for high-angle-of-attack and spin characteristics of the YF-17 prompted the Navy to request the full suite of tests at Langley for these characteristics. A request for tests in the Spin Tunnel and the Full-Scale Tunnel and helicopter drop models was received in late 1976.

The first preproduction F/A-18 made its first flight on November 18, 1978, and entered the initial phases of flight tests at the Patuxent River Naval Air Station in Maryland. The preproduction flight-test program lasted from January 1979 to October 1982, and the Langley staff was called on to help solve several critical developmental problems.

Cruise Drag

Initial results from flight evaluations at Patuxent River in 1979 indicated that the cruise performance of the F/A-18 was significantly below expectations, with a shortfall of about 12 percent in cruise range. The performance deficiency became a weapon for those who sought the termination of the F/A-18 Program. A number of reasons for the poor performance were identified. Modifications to the engines, computer-controlled schedules for the deflection of leading- and trailing-edge flaps, and other changes reduced the cruise range deficit to about 8 percent, but aerodynamic drag remained a problem.

In response to this critical threat to the program, a Navy and NASA F/A-18 working group was formed in late 1979. The NASA members were all Langley personnel led by researchers Richard Whitcomb, Edward Polhamus, and William J. Alford, Jr. With the addition of members from the Navy, McDonnell Douglas, and Northrop, the group totaled about 20 participants. After consideration of several approaches to reduce the drag of the aircraft, the group recommended wind-tunnel and flight studies of modifications to several configuration features. Modifications included increasing the wing leading-edge radius, variations in the LEX camber, and filling in the slots in the LEX-fuselage juncture. The Langley members identified the slots as a particularly undesirable feature with potentially high drag characteristics. Tests with favorable results were conducted in the Langley 8-Foot Transonic Pressure Tunnel in early 1980. These changes were implemented on the F/A-18 test aircraft at Patuxent River where they were found to favorably increase the cruise range of the aircraft. The impact of filling in the LEX slot on high-angle-of-attack characteristics was found to be acceptable in additional tests at Langley and F/A-18 flight tests.

High-Angle-Of-Attack, Spin, and Spin Recovery Characteristics

Langley responded to the Navy's request for stall-spin tests in support of the F/A-18 with spin tunnel tests (1978), free-flight model tests (1978), and drop-model tests (1979). The results of the Langley spin tunnel and drop-model tests were very favorable. The F/A-18 configuration was found to be extremely resistant to spins. (The pilot was required to maintain prospin controls for over 20 sec to promote a spin.) When spins were entered, recovery could be effected very quickly. In the spin tunnel tests, the F/A-18 model demonstrated the best spin recovery characteristics of any modern U. S. fighter (as had the YF-17 configuration). During the limited model tests for spins, the phenomenon known as "falling leaf" was not encountered, but it became a problem in operational usage as will be discussed in a later section.

The F/A-18 free-flight model tests that were conducted in the Full-Scale Tunnel were very controversial; however, the model tests subsequently proved to be a major contributor to the success of the F/A-18 development program.

As previously discussed, early high-angle-of-attack tests had been completed in other NASA and McDonnell Douglas wind tunnels to tailor the geometry of the F/A-18 for good flight characteristics. The removal of the stabilizing nose strakes that were on the YF-17, which would have interfered with the radar performance on the F/A-18, and the revision of LEX shape had been carefully analyzed and designed from these results. When the Langley free-flight model underwent tests, the LEX and leading-edge flap schedule had been defined for flight tests.

When the Langley model was tested, however, the results obtained with this particular model indicated that the F/A-18 would exhibit a moderate yaw departure near maximum lift. Although not a flight safety concern, this result would infringe on the precision maneuverability of the aircraft. These undesirable results had been obtained at low wind-tunnel test speeds with a relatively large model and were dismissed with skepticism by the engineering community as having been caused by erroneous scale effects.

In 1979, an F/A-18 test aircraft at Patuxent River suddenly and unexpectedly departed controlled flight during a wind-up turn maneuver at high subsonic speeds. None of the baseline wind-tunnel data predicted this characteristic, and the F/A-18 Program was shocked by the event. The fact that the free-flight model had also exhibited such a trend did not go unnoticed, and a joint NASA, Navy, and McDonnell Douglas team was formed to seek solutions with the free-flight model at Langley. Following exhaustive wind-tunnel tests in the Full-Scale Tunnel, the team recommended that the wing leading-edge flap deflection be increased from 25 deg to 34 deg at high angles of attack. Following the implementation of this recommendation on the test aircraft (via the flight

F/A-18 drop model prepared for flight in 1978.

The F/A-18A free-flight model during tests in the Langley 30- by 60-Foot Wind Tunnel in 1978.

control computers), no more departures were experienced, and the flap deflection schedule was adopted for production F/A-18's.

This was not the first time that results from large free-flight models in the Full-Scale Tunnel had proven to accurately predict the flight behavior of an aircraft at high angles of attack, despite the low speeds of the wind-tunnel tests. (See *Langley Contributions to the F-16*, for example.)

The NASA High-Angle-Of-Attack Technology Program

The emergence of a generation of highly maneuverable fighter aircraft such as the F-14, F-15, F-16, and F-18 in the late 1970's resulted in a new perspective on operating high-performance aircraft at high angles of attack. Previous U.S. military experience with aircraft such as the F-4 and A-7 during the Vietnam era had been tormented with unacceptably high accident losses of these aircraft from inadvertent departures and spins from maneuvers at high angles of attack. Operational procedures required very careful and precise pilot inputs at high angles of attack to avoid loss of control, and handbook restrictions were placed on the operational use of angle of attack. With the advent of the new generation of fighters, flight at high angles of attack became a common occurrence and was no longer feared or avoided by pilots. Exploitation of high angles of attack provided the potential for new maneuvers and options for air combat tactics. These interests resulted in several major programs within DOD, industry, and NASA to develop the analysis and design methodologies for superior performance over an unlimited range of angle of attack.

In the early 1980's, NASA initiated a High-Angle-of-Attack Technology Program (HATP) among the aeronautics research centers (Langley, Ames, Dryden, and Glenn). The program included all the critical elements of high-angle-of-attack technology: aerodynamics, flight controls, handling qualities, stability and control, and the rapidly evolving area of thrust vectoring. In recognition of its extensive accomplishments in this field, Langley was designated the technology lead center for the program, and Dryden was designated the lead for flight research and operations. The Langley leader in the initial phases of the program was William Gilbert, who was followed by Luat Nguyen in later years.

A critical element in the HATP plan was the correlation and validation of experimental and analytical results with results from flight tests of a high-performance aircraft. The NASA team considered several aircraft configurations for this important task, including the F-15, F-16, F/A-18, and the X-29 forward-swept wing technology demonstrator aircraft. The advantages and disadvantages of each aircraft for the NASA program were thoroughly considered, and the F/A-18 was unanimously chosen as the desired test vehicle for several reasons. In flight, the F/A-18 had shown aerodynamic and aeroelastic phenomena of interest (vortex flows and tail buffet), stall-free engine operation at high angles of attack, excellent spin recovery characteristics, an advanced digital flight control system, and a large angle-of-attack capability (up to 60-deg trim capability at low speeds). The Navy's response to a request from NASA for a preproduction F/A-18 vehicle for the program was very positive. The Navy initially offered NASA an early production F/A-18A aircraft; however, NASA targeted the specially equipped preproduction F/A-18 (ship no. 6) that had been used at Patuxent River for the spin evaluation phase of the development program. The aircraft had completed its spin tests and had been stored in a hangar and stripped of its major components as a spare-parts aircraft. However, this particular aircraft was equipped with a special emergency spin recovery parachute system worth several million dollars and a programmable digital flight control computer adaptable to the variations desired in the HATP. The Navy approved the transfer of the stripped aircraft to NASA, and it was trucked to Dryden and reassembled by experts at Dryden with the Navy's help into a world-class research aircraft to be known as the F/A-18 High Alpha Research Vehicle (HARV).

F-18 HARV during flow visualization studies using smoke
ejected at apex of LEX and wool tufts on upper surfaces.

The HATP studies were conducted in three phases. In the first phase, emphasis was placed on static and dynamic aerodynamic phenomena. This phase included extensive correlations of wind-tunnel and computational results with data from flight instrumentation on the HARV. These data correlations were conducted with common data sensor locations on the wind-tunnel models and the HARV. As a result of these aerodynamic experiments, the capabilities of emerging computational fluid dynamics tools and the interpretation of wind-tunnel test techniques were aggressively accelerated. Detailed studies of vortical flow structures emanating from the F/A-18 LEX and interactions with the vertical tail surfaces resulted in rapid progress in the understanding, prediction, and minimization of vertical-tail buffet phenomena. Finally, in-depth analysis of the flow fields shed by the nose and forebody of the F/A-18 resulted in wind-tunnel test procedures and validation of computational codes for future fighters.

An excellent example of the impact of this fundamental research is the LEX upper-surface fences mounted on F/A-18C/D aircraft. These devices greatly alleviate the vertical tail buffeting associated with sudden bursting of the core of the strong vortical flow shed by the LEX at high angles of attack. McDonnell Douglas developed the fence during parametric wind-tunnel experiments in its own wind tunnel; however, they did not have the resources or time to identify the flow mechanisms created by the fence. Within the scope of the HATP, Langley developed a laser vapor screen flow visualization technique that permitted rapid, global assessments of interactive flow fields. Using this technique, Langley researcher Gary E. Erickson conducted diagnostic tests on an F/A-18C model in the Navy David Taylor Research Center (DTRC) 7- by 10-Foot Transonic Tunnel, and he precisely identified the vortex-flow mechanisms associated with the benefits of the fences. The success of the vapor screen system so impressed the Navy and industry that they requested the Langley vapor screen system on other tests to understand nonlinear flow behavior at subsonic and transonic speeds.

In the second phase of the HATP, the HARV incorporated a relatively simple and cheap thrust-vectoring system for studies of aerodynamics at extreme angles of attack, and engineering development and evaluations of the advantages of thrust vectoring for high-angle-of-attack maneuvers. Previous efforts with the Navy involving experiments with thrust-vectoring vanes on an F-14 inspired the NASA team to adopt this simple concept instead of a program to develop internal engine vectoring at a cost of many more millions of dollars. (See *Langley Contributions to the F-14.*) Dryden managed and directed a contract with McDonnell Douglas to outfit the HARV with deflectable external vanes mounted behind the engines. The engine exhaust nozzle divergent flaps were removed and replaced with a set of three vanes for each engine, which resulted in both pitch and yaw vectoring capability. The specific vane configuration was analyzed and defined by tests of a powered model in the Langley 16-Foot Transonic Tunnel and the Langley Jet Exit Test Facility. Meanwhile, the aircraft flight computer was modified to permit research evaluations within a broad spectrum of thrust-vectoring parameters. The modification of the HARV proved to be extremely challenging, and several members of the staffs from Langley and Dryden assisted in the final implementation of the hardware and software of the thrust-vectoring system.

Extensive piloted simulator evaluations of the HARV with thrust vectoring were conducted in the DMS, with results indicating that the implementation of both pitch and yaw vectoring provided powerful, unprecedented controllability and precision for maneuvers at high angles of attack.

The results of the F/A-18 HARV thrust-vectoring flights were remarkable. The HARV became the first high-performance aircraft to conduct multiaxis thrust-vectoring flights,

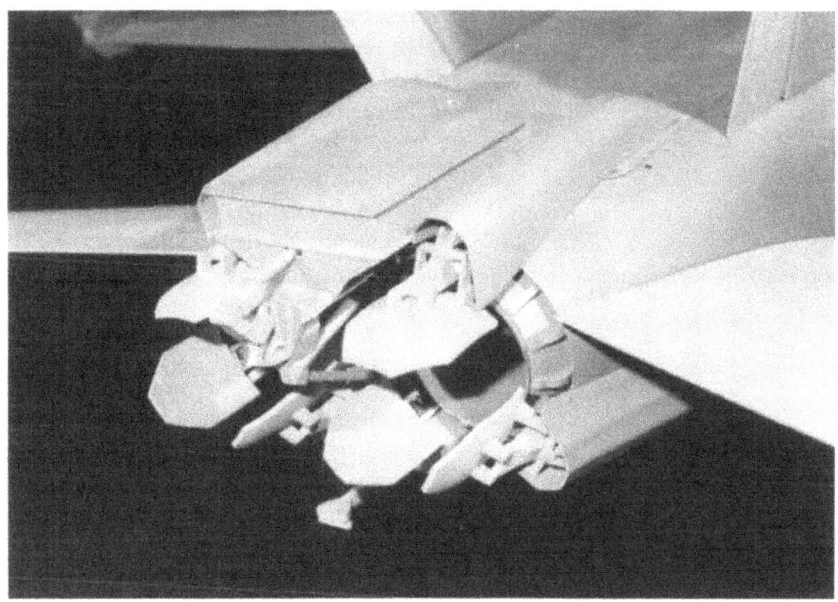

Thrust-vectoring vane installation on the F-18 HARV free-flight model with rectangular box above the vanes representing the container for the spin parachute.

and the powerful effectiveness of thrust vectoring at high angles of attack was vividly demonstrated. The control precision of the aircraft at extreme angles of attack was significantly enhanced, and the vehicle could be stabilized at the conditions for high quality aerodynamic studies.

In the final phase of the HATP, earlier studies to understand and predict the strong vortex flows emanating from the long pointed fuselage forebody of the F/A-18 were extended. The studies now included the ability to control these flows for aircraft maneuvers at extreme angles of attack, where yaw control provided by conventional aerodynamic rudders becomes relatively ineffective. Based on in-depth studies conducted by Langley in several wind tunnels, the HARV was equipped with deflectable foldout strake surfaces on the forward forebody, and the control system was modified to permit the pilot to use the strakes for roll control. The strake hardware was engineered and fabricated in Langley machine shops, and the flight computer interface was provided by Dryden. Results of the flight evaluation of the nose strakes were extremely impressive. The time required to roll the aircraft through a specific roll attitude change was significantly reduced due to the crisp, powerful control provided by the strakes. At the higher subsonic speeds, the effectiveness of the strakes approached levels provided by thrust vectoring in yaw.

The HATP efforts represent a decade of invaluable research and development on key technologies that will be required for future highly maneuverable high-performance fighter aircraft, as well as the F/A-18 configuration. The data gathered by NASA has been widely disseminated among design teams at NASA sponsored national conferences and during site visits. The program has been cited for its high quality, pioneering value on numerous occasions by industry and DOD. The results of the program are being used in the development of the next generation F/A-18E/F and other military aircraft.

The Falling-Leaf Maneuver

The falling-leaf maneuver originated during World War I as a flight training exercise. In this exercise, pilots intentionally stalled the aircraft and forced a series of incipient spins to the right and left. The aircraft descends as it rocks back and forth, much as a leaf does falling to the ground. In the early 1980's, an unintentional falling-leaf mode surfaced as a severe out-of-control problem during developmental flight tests of the F/A-18A. The out-of-control falling-leaf mode is a highly dynamic mode where the aircraft oscillates so that it is very difficult to reduce angle of attack and recover. The term "alpha hang-up" was used to describe this problem with the F/A-18 and it was a key driver in establishing the aft center of gravity and the maneuvering limits for the aircraft. During early operational use of the F/A-18, the falling-leaf mode was rarely encountered; however, by the early 1990's increasingly aggressive maneuvering had exposed a susceptibility to the falling-leaf mode with numerous incidents and losses of aircraft.

Langley began studying the falling-leaf mode in 1994 at the request of the Naval Air Systems Command, following an F/A-18 falling-leaf incident during a routine training flight that nearly resulted in loss of the aircraft. Under the technical leadership of Langley researcher John V. Foster, the cause for the falling-leaf mode was identified and linked to a common aerodynamic stability design method that had been used for many years. Using the DMS, the critical entry maneuver that excites the mode was identified and correlated with fleet incidents, and various recovery methods were studied. In addition, the necessary requirements for high-fidelity prediction of the falling-leaf motion in piloted real-time simulations were defined.

In light of the growing falling-leaf problem on early models of the F/A-18, there was concern that the emerging F/A-18E/F, which was then preparing for developmental flight tests, would have the same problem. In response, McDonnell Douglas proposed a control law design approach, supported by Foster's research, that specifically targeted falling-leaf suppression, which was later implemented on the aircraft.

Because no methods existed specifically for falling-leaf flight tests, NASA again used the DMS facility to develop reliable flight-test techniques that were subsequently validated on the NASA F/A-18 HARV at Dryden. Using these test techniques, falling-leaf susceptibility was extensively evaluated on the F/A-18E/F during the high-angle-of-attack flight-test program. While the unaugmented aircraft was shown to exhibit the falling-leaf mode, the new control system design was shown to be very effective in suppressing the mode and the falling-leaf problem was considered solved for that aircraft. Largely due to the success of the F/A-18E/F program, the Navy is considering retrofitting earlier models of the F/A-18 with the updated control law for the purpose of eliminating the falling-leaf problem.

While the F/A-18 was the most visible recent manifestation of the falling-leaf problem, numerous other aircraft configurations have exhibited the mode. For example, the AV-8B exhibited a severe falling-leaf mode that the Foster criteria showed to be a result of a minor configuration change. As new highly maneuverable configurations are developed and the angle-of-attack envelope expands, Langley's F/A-18 research has brought a necessary and timely focus to a critical flight problem.

Flutter Clearance and Tail Buffet

In a departure from most fighter programs, flutter clearance tests of both the YF-17 and the F/A-18A were not conducted in the Langley 16-Foot Transonic Dynamics Tunnel. Research studies were, however, conducted in the areas of active flutter suppression and buffet alleviation. A cooperative NASA, Northrop, and Air Force research study was later conducted to assess the application of active controls for flutter suppression for the

YF-17 wing-store-pylon combination in the tunnel in late 1977. The results indicated that active control of wing-store-pylon flutter was possible and that wing leading-edge surfaces could be effective in such a control system. In 1979 interest in the concept resulted in an international study of several active flutter suppression systems designed by British Aerospace and the Royal Aeronautical Establishment from the United Kingdom, the Office of National d'Etudes et de Recherches Aerospatiales from France, and Messerschmitt-Bolkow-Blohm from the Federal Republic of (West) Germany. The TDT tests showed that all the control systems were highly effective in supressing flutter. The YF-17 testing continued through 1982.

The vertical-tail buffet experienced by the F/A-18A and other high-performance aircraft resulted in cooperative research by industry, DOD, and NASA on buffet alleviation from smart materials and actively controlled rudders. Because international allies of the U.S. use the F/A-18, this research effort included Australia and Canada. The Actively Controlled Response of Buffet Affected Tails (ACROBAT) Program led by Robert W. Moses investigated the use of an actively controlled rudder, actively controlled piezoelectric devices, and other novel concepts to alleviate vertical-tail buffeting. For the ACROBAT Program, a 1/6-scale rigid full-span model of the F/A-18 A/B aircraft with flexible tails was tested in the 16-Foot Transonic Dynamics Tunnel. Numerous control laws were tested, and many demonstrated buffeting alleviation over a large angle-of-attack range. These tests have been followed up with international ground-based tests of a buffeting alleviation system on a full-scale F/A-18.

Thrust-Vectoring Research

In the early 1970's, the Navy conducted a vertical and short takeoff and landing (V/STOL) nozzle study for potential applications to high-performance aircraft. Following this study, the augmented deflector exhaust nozzle (ADEN) was selected as the best nozzle concept for Navy V/STOL requirements. The ADEN nozzle had a hood, which deployed from a long upper ramp, that was capable of thrust vector angles from 0 deg to over 90 deg. Nozzles of this type (long flap on one side of the nozzle and no deploying hood without VTOL capability) were later called single expansion ramp nozzles (SERN).

F/A-18 model with 2-D convergent-divergent (CD) nozzles tested in Langley 16-Foot Transonic Tunnel.

In 1975, a joint DOD and NASA two-dimensional (2-D) nozzle workshop held at Langley formed an ad hoc interagency nonaxisymmetric nozzle working group that eventually recommended a flight test of 2-D, thrust-vectoring, in-flight thrust-reversing nozzles. To avoid duplication, studies on potential flight-test vehicles were split up among NASA (F-15), the Air Force (F-111), and the Navy (YF-17 and F/A-18).

The first full-scale tests, which were sponsored by the Navy, of a 2-D nozzle (ADEN) were conducted in a Glenn Research Center altitude test cell in 1976. In support of the Navy studies of advanced nozzles, Langley conducted powered-model tests of an early F/A-18A configuration with 2-D nozzles in the Langley 16-Foot Transonic Tunnel.

Since the F/A-18A represented the state of the art in fighter design, Langley expanded the test objectives to include propulsion-airframe integration studies of a number of nozzle types (wedge, 2-D convergent-divergent (CD), SERN), thrust-vectoring concepts, and in-flight thrust-reversing concepts. The studies also examined F/A-18A vertical-tail loads, particularly during in-flight thrust-reversing operation. Unfortunately, Navy and NASA attempts to secure funding for the F/A-18A flight tests were unsuccessful, and the F/A-18A studies of thrust vectoring were terminated.

F/A-18E/F

The Super Hornet

The F/A-18E/F Super Hornet was funded in June 1992 as the replacement for the cancelled Navy A-12 aircraft and as the replacement for the early F/A-18A aircraft and other Navy and Marine Corps aircraft. The F/A-18E/F is a larger version of the F/A-18C/D Hornet with extended mission capabilities. The E/F is roughly 25 percent larger than the C/D, with a 25 percent increase in operating radius and a 22 percent increase in weapons load capability. First flight of the F/A-18E/F occurred on November 29, 1995, and preproduction flight tests began in February 1996, at Patuxent River.

F/A-18E flight test aircraft prepares to land at Patuxent River Naval Air Station.

Cruise Performance

NASA's involvement with the F/A-18E/F began in the early stages of the proposed aircraft development when the Office of the Secretary of Defense (OSD) became concerned with the accuracy of McDonnell Douglas' range estimates for the vehicle. At the request of OSD, a three-member NASA, DOD, and industry team conducted an independent review of McDonnell Douglas' F/A-18E/F fighter-escort mission range estimates in April 1992. Langley researcher Gary Erickson represented NASA and called on supporting analyses from several Langley specialists. Results of the independent review substantiated the McDonnell Douglas estimates and were briefed to the Assistant Secretary of Defense. The favorable results from this review were critical to the aircraft program proceeding to the Defense Acquisition Board for funding advocacy.

Redesign of the Leading-Edge Extension

Many operational requirements were revisited as the F/A-18C/D grew to the larger, heavier F/A-18E/F. Critical performance issues included lift requirements for carrier approach and wave offs, high-angle-of-attack stability, recovery from extreme attitudes, and minimization of adverse aerodynamic phenomena such as tail buffet. McDonnell Douglas explored a number of configuration changes to satisfy these requirements, including adding a snag to the wing leading edge, reshaping the LEX, and using moveable spoilers on the upper surfaces of the LEX to ensure recovery from extreme angles of attack.

Low-speed wind-tunnel tests by McDonnell Douglas had indicated that a deflected spoiler on the upper surface of each LEX would alleviate buffeting of the vertical tails, promote good lateral-directional stability, and provide trim changes for nose-down control at extreme angles of attack. In May 1992, a series of tests were conducted in the Langley 8-Foot Transonic Pressure Tunnel to determine the effects of the spoiler concept at higher speeds (Mach numbers from 0.6 to 1.2). Unfortunately, results of the

F/A-18E/F model in 8-Foot Transonic Pressure Tunnel for evaluations of the LEX spoiler concept

F/A-18E/F (foreground) with redesigned LEX in flight with F/A-18C chase aircraft.

Langley tests indicated that the deployed spoilers would cause an unacceptable decrease in maximum lift of over 15 percent. With these negative results, a redesign of the LEX surfaces began. The LEX spoilers have been retained on the F/A-18E/F as speed brakes and nose-down control devices.

Redesigning the LEX was the job of a 15-member industry, DOD, and NASA team, which included Langley researchers Daniel G. Murri, Robert M. Hall, and Gary Erickson. This team, which was active for the first 6 months of 1993, initially explored small modifications to the size and shape of the original F/A-18C LEX to help provide the required lift and improve lateral-directional stability. However, subsequent wind-tunnel tests showed that this incremental approach would not be successful and much larger changes to the LEX configuration would be required. Based on his prior research with other configurations, Murri proposed more radical LEX candidates that would potentially satisfy these requirements. One of the LEX configurations with favorable wind-tunnel results that was recommended by Langley was accepted for further refinements and met all design goals. This configuration was the basis for the final design adopted as the wing LEX configuration for the production F/A-18E/F.

Extensive tests of the F/A-18E/F with the redesigned LEX were conducted in several NASA and industry wind tunnels to completely define the aerodynamic characteristics of the configuration. Data from these tests were used to generate the aerodynamic databases for flight simulation and to develop the flight control software for the aircraft. NASA engineers worked closely with engineers from McDonnell Douglas and the Navy to assure that design requirements and national goals were met.

High-Angle-Of-Attack Characteristics

Stability and control characteristics of the F/A-18E/F at high-angle-of-attack flight conditions were evaluated in numerous wind-tunnel tests at Langley. In the Full-Scale Tunnel, a combination of static, dynamic, and powered tests was conducted to define and develop a database for the high-angle-of-attack aerodynamic, stability, and control characteristics of the aircraft. In addition to tests of the basic configuration, tests were conducted to study the impact of fuselage-mounted and wing-mounted stores on aerodynamic and stability characteristics, to assess aerodynamic damping characteristics, and to assess the magnitude of thrust-induced aerodynamic effects on the configuration. Free-flight tests were also conducted to provide confirmation of the stability and flight dynamic characteristics.

This database has been used by McDonnell Douglas and the follow-on Boeing organization, NASA, and the Navy to conduct flight-simulation studies and to aid in the development of the aircraft's flight control system. An extensive unpiloted and piloted simulation of the F/A-18E/F has been implemented in the Langley DMS and is being used to analyze flight control modifications and the impact of aircraft parameters on high-angle-of-attack behavior.

Spin Tunnel and Drop-Model Tests

Hundreds of tests in the Langley Spin Tunnel quantified the spin modes and spin recovery characteristics of the F/A-18E/F, determined the acceptable emergency spin recovery parachute size, and identified the optimal spin recovery procedures prior to flight tests. Time histories of model motions from these tests were used by McDonnell Douglas to validate their spin simulation. Data from rotary-balance tests conducted in this facility provided inputs for an analytical assessment of spin modes, spin recovery characteristics, and a database for incorporating rotational aerodynamic characteristics into the flight simulation.

F/A-18E/F drop model immediately after release from the helicopter.

A highly sophisticated F/A-18E/F drop model was tested by Langley at the NASA Wallops Flight Facility to provide risk reduction for the high-angle-of-attack part of the flight-test program. These tests, which used a 0.22-scale remotely piloted model, supplemented the aircraft flight-test program by providing flight dynamics data for the aircraft at conditions outside the planned operating envelope.

Flutter Tests

F/A-18E/F flutter clearance tests were conducted at Langley by Moses G. Farmer in the 16-Foot Transonic Dynamics Tunnel during four tunnel entries from 1993 to 1995. Phase I tests insured that each pair of dynamically scaled surfaces (wings, horizontal tails, and vertical tails) was clear of flutter throughout the scaled flight envelope. Phase II of the tests studied the configuration both with and without stores (bombs and fuel tanks) mounted on the wings. These tests used the unique two-cable-mount system of the tunnel, which allows the model to actually fly in the center of the tunnel with assistance from a pilot in the control room. The tests verified that the aircraft was free from aeroelastic instabilities, including flutter, within its flight envelope. As a result of the Langley tests, the number of expensive flight tests dedicated to flutter clearance could be minimized.

In view of Langley's corporate knowledge of the F/A-18E/F and its predecessors, the Navy requested NASA involvement when the F/A-18E/F began flight tests. NASA continued to work closely with the Navy and McDonnell Douglas during the engineering and manufacturing development (EMD) phase. Langley support included flight-test planning and data evaluation, especially in the high-angle-of-attack regime. In 1997, Langley assigned an engineer to Patuxent River as a participant in the flight tests and as a liaison and technical consultant for the test team.

F/A-18E/F model mounted for flutter tests in the Langley 16-Foot Transonic Dynamics Tunnel.

Wing Drop

The most recent Langley contribution to the F/A-18E/F Program explored the phenomenon known as wing drop. In March 1996, during flight tests at Patuxent River, an F/A-18E/F experienced wing drop—an unacceptable, uncommanded abrupt lateral roll-off that randomly occurred and involved rapid bank angle changes of up to 60 deg. The problem was viewed as extremely serious and posed a threat to operational tests and the overall development schedule. During the first year of flight tests, the wing-drop phenomenon was only seen at high altitudes because load restrictions prevented the aircraft from reaching the relevant range of angle of attack at low altitudes. As the loads test program opened the flight envelope to 7.5g at all altitudes, the full extent of the wing-drop problem became evident. Objectionable wing-drop events occurred throughout the flight envelope at Mach numbers between 0.5 and 0.9, and this deficiency became a significant threat to the technical and political health of the F/A-18E/F Program.

A joint Navy and McDonnell Douglas team concluded that the wing drop was caused by a sudden, abrupt loss of lift on one of the outer wing panels during maneuvering. Though the basic cause of the wing drop was determined, how to moderate the airflow separation differences between the left and right wings was not. A variety of solutions was explored. For example, 25 potential wing modifications were tested in wind tunnels, computational fluid dynamics studies were undertaken, and two of the potential fixes were flown with mixed results. In one approach, the use of stall strips in the vicinity of the wing-fold fairing eliminated wing drop, and the aircraft exhibited excellent flight qualities throughout the envelope. However, the impact of the stall strips on range and turn performance were severe, and they were discarded as a viable option.

Langley engineer Robert Hall also served on a DOD blue ribbon panel to review the approach taken by McDonnell Douglas to resolve the wing drop. The panel also participated on various McDonnell Douglas and Navy "tiger teams" that were created to resolve issues related to the wing-drop problem. Roy V. Harris, a distinguished Langley retiree, was selected by DOD to head the blue ribbon panel, which subsequently cited the lack of technical understanding and design tools available in the technical community to address the high subsonic and transonic wing-drop problem. The panel strongly recommended that a research program be undertaken to develop the design methods required to avoid this problem in future fighter aircraft. The recommendation was accepted, and a joint NASA and Navy Abrupt Wing Stall (AWS) Program was initiated. Langley researchers Jeffrey A. Yetter and Robert Hall served as AWS Program Comanager and Technical Comanager, respectively, with their Navy partners.

As a low impact "80-percent solution" to the F/A-18E/F wing-drop problem, a revised deflection schedule for the leading-edge flaps was evaluated in flight tests in early 1997, with very favorable results. Although the leading-edge flap schedule modification significantly reduced the magnitude of the problem, the aircraft still exhibited smaller wing drops at many test conditions. The original wing-drop problem had been viewed as a potential safety hazard and a roadblock to productive load tests. After the modification, the problem was reduced to a flying qualities issue that allowed other tests to continue. Wind-tunnel tests were conducted at sites other than NASA to evaluate wing changes that might eliminate the residual wing-drop tendencies. Langley researchers Gary Erickson and Robert Hall participated in the tests, which continued through the summer of 1997.

During the winter of 1997 to 1998, the Navy asked NASA for additional assistance in resolving wing drop. In November 1997, a diagnostic flight test of an F/A-18E/F with the wing-fold fairing removed was conducted and the results showed that wing drop had been eliminated from most of the flight envelope. The fairing-off configuration was not a viable approach for production aircraft, and Langley engineers led by Steven X. S. Bauer suggested that the flight program apply porosity, a passive technology developed at NASA, to the wing in the fold area. Langley researchers had been conducting experiments with passive porosity to control shock locations and other characteristics for several years. During the initial concept evaluation on the F/A-18E/F, the porous fairing was simply a standard wing-fold fairing with areas cut out and a screen mesh substituted. Langley provided design guidelines for the porosity and thickness of the mesh. This solution, refined by the NASA, Navy, and Boeing team, resolved the wing-drop problem and permitted continued production of the aircraft.

Additional activities are being conducted to support the joint NASA and Navy AWS Program, including tests of an F/A-18E/F model in the Langley 16-Foot Transonic Tunnel in September 1999 under the leadership of Robert Hall and S. Naomi McMillin.

Boeing T-45 Goshawk

SPECIFICATIONS

Manufacturer
  Boeing
Date in service
  January 1992
Type
  Trainer
Crew
  Two
Engine
  Rolls Royce F405-RR-401

USER

U.S. Navy

DIMENSIONS

Wingspan . . . . 30.1 ft
Length . . . . 39.3 ft
Height . . . . 14.0 ft
Wing area . . . . 180 sq ft

WEIGHT

Empty . . . . 9,394 lb
Gross . . . . 13,636 lb

PERFORMANCE

Max speed . . . . . Mach number
                            of 1.04

HIGHLIGHTS OF RESEARCH BY LANGLEY FOR THE T-45

1. At the request of the Navy, Langley participated in an independent assessment team review of technical progress during early development.

2. Langley conducted exploratory research to define potential wing leading-edge modifications to cure severe low-speed wing drop on the original configuration.

3. Tests were conducted at Langley to determine the spin and spin recovery characteristics of the aircraft and its potential utility as a spin trainer.

4. Langley provided critical tire characterization data that dramatically improved the modeling and analysis of pilot-induced oscillations during landing.

5. Tests were conducted in the Langley 16-Foot Transonic Tunnel to identify potential candidates for a revised engine inlet configuration.

The Boeing (formerly McDonnell Douglas) T-45 Training System is the first totally integrated training system developed for and used by the U.S. Navy. It includes the T-45 Goshawk aircraft built by Boeing, advanced flight simulators, computer-assisted instructional programs, a computerized training integration system, and a contractor logistics support package. The integration of all five elements produced a superior pilot in less time and at lower cost than previous training systems. The two-seat, single-engine T-45 Goshawk is the heart of the training system. The T-45 replaces the T-2 and TA-4 aircraft formerly used for Navy advanced jet trainers. The Goshawk's design is based on the British Aerospace Hawk land-based aircraft; however, design modifications have been made to the Goshawk that make the aircraft more suitable for carrier-based operations. The modifications include strengthened landing gear, the addition of an arresting hook, and catapult launch fittings. The latest version of the aircraft, known as the T-45C, includes a digital cockpit.

Langley's involvement in the T-45 began with an independent assessment team review of technical progress during the initial development program in 1988. Subsequent activities included spin tunnel tests, wind-tunnel research on wing leading-edge modifications, characterization of landing gear tires for improved handling characteristics during ground rollouts on landing, and wind-tunnel assessments of new engine inlet designs.

Langley facilities involved in support of the T-45 Program included the 16-Foot Transonic Tunnel, the 12-Foot Low-Speed Tunnel, the 20-Foot Vertical Spin Tunnel, and the Landing Loads Facility.

LANGLEY CONTRIBUTIONS TO THE T-45

Early Configuration Development

The U.S. Navy initiated the Advanced Trainer (VTX) Program in the early 1980's to replace the existing T-2 Buckeye and TA-4 Skyhawk advanced jet trainers. Industry responses to the Navy request for proposals (RFP) included several existing and new aircraft configurations. A team from McDonnell Douglas and British Aerospace proposed both a modification of the existing British Hawk land-based configuration and a new trainer. The VTX contract was awarded to the McDonnell Douglas and British Aerospace team in November 1981. The Boeing (formerly McDonnell Douglas) T-45 Goshawk evolved from the Hawk design.

Conversion of the Hawk land-based aircraft to a naval trainer with carrier capabilities involved considerable research and development. In addition to the necessary strengthening of landing gear components and the inclusion of arresting gear, development work was required in numerous areas that were critical for carrier-based operations. Some areas of concern included the handling qualities, engine response characteristics, and stall characteristics of the T-45.

In 1988, following extensive preliminary flight-test evaluations by the Navy at the Patuxent River Naval Air Station in Maryland, the Navy cited several major deficiencies in the T-45. The deficiencies included high approach speed, slow engine thrust response, and longitudinal and lateral stability deficiencies. McDonnell Douglas and British Aerospace developed candidate solutions and recommended approaches to resolve these issues. In 1989 the Navy convened a blue ribbon independent assessment team to review the technical status and plans for the program. In response to a request from the Navy for Langley representation, Joseph R. Chambers served on the team.

Stall Characteristics

The stall characteristics of the initial T-45 configuration were judged to be unacceptable by the Navy on the basis of a severe wing-drop behavior at the stall and high approach speeds (aggravated by the increased weight required to strengthen the airframe for

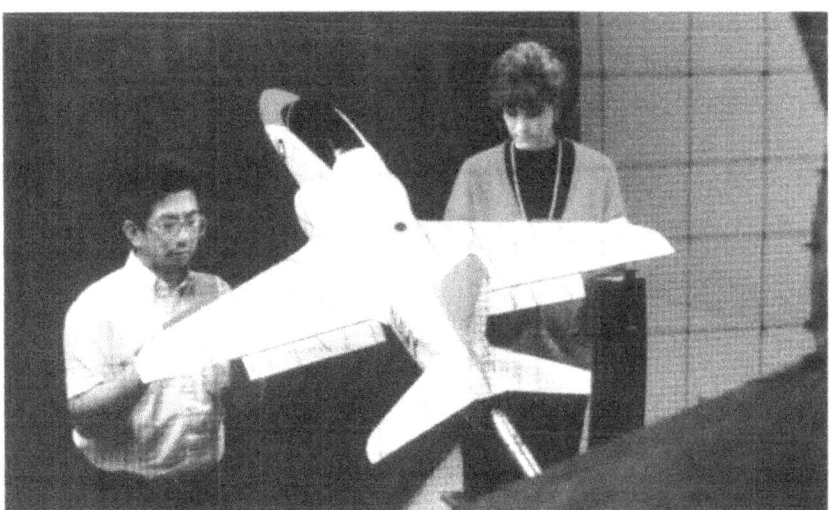

Langley researchers Long P. Yip and Holly M. Ross examine a T-45 model with a discontinuous leading-edge modification in the Langley 12-Foot Low-Speed Tunnel in 1989.

carrier operations). During the Navy's flight evaluations, the wing drop was so severe that uncommanded roll motions often exceeded 90 deg.

While McDonnell Douglas and British Aerospace explored various candidate solutions to the wing-drop issue, Langley proposed a joint research project to assess the effectiveness of a Langley concept for enhanced stall characteristics. As part of research to improve stall and spin characteristics of general aviation aircraft, Langley had conceived a wing leading-edge modification that significantly improved the lateral stability and spin resistance of several general aviation research aircraft during stall-spin research flight tests. The leading-edge modification consisted of a discontinuous "snag" at about the 75-percent span location that acted similar to an aerodynamic fence and maintained attached flow on the outer wing panel to very high angles of attack (beyond stall). Although the wing leading-edge radius of the T-45 was much smaller than those of general aviation aircraft, it was hoped that the modification would also be effective for the advanced trainer.

Led by Langley's Long P. Yip, tests were conducted on a 0.15-scale T-45 model in the Langley 12-Foot Low-Speed Tunnel. The test techniques included conventional static force tests as well as single-degree-of-freedom free-to-roll tests to determine wing-drop tendencies. The results of the investigation indicated that the wing modification was effective in minimizing wing-drop tendencies for the cruise configuration; however, the modification was not effective for the landing configuration with the wing trailing-edge flaps deflected. The T-45 Program subsequently adopted a wing redesign, which incorporated wing leading-edge slats. The slats virtually eliminated the wing-drop tendency and lowered the carrier-approach speed to a more acceptable value.

Spin Characteristics

Flight-test experience with the British Hawk aircraft had indicated that the aircraft was very reluctant to spin and that attempts to intentionally spin the aircraft usually resulted

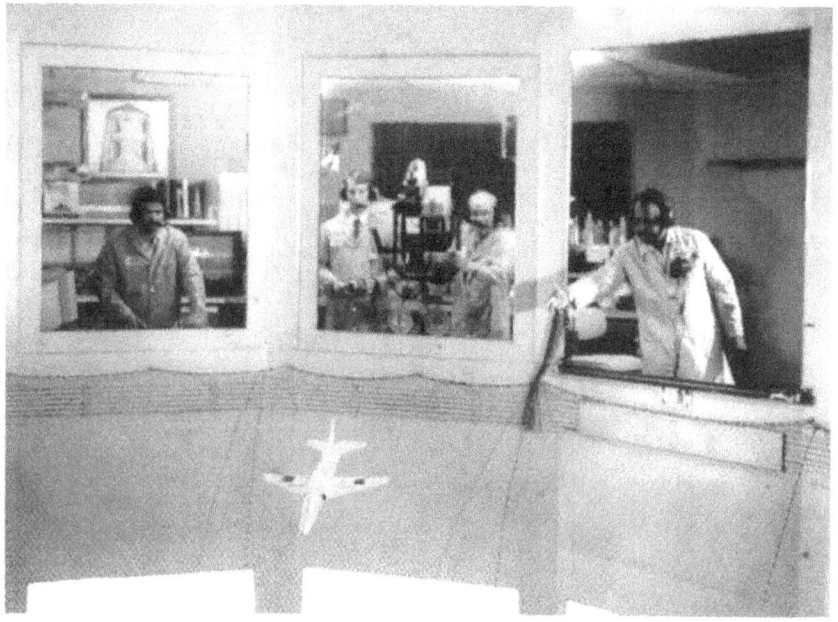

T-45 model undergoing test in the Langley Spin Tunnel.

in a spiral with rapidly increasing airspeed. The Hawk recovers from the high-speed spiral when controls are neutralized. Spin tests of the T-45 in the Langley 20-Foot Vertical Spin Tunnel by Langley researcher Raymond D. Whipple indicated that the Goshawk would exhibit similar characteristics. In fact, it was virtually impossible to obtain spins for the T-45 model in the normal upright attitude because, after being hand launched with prerotation into the vertical airstream, the model would immediately nose over and rapidly increase airspeed beyond the test capability of the Spin Tunnel.

Full-scale flight tests of the T-45 subsequently verified the predictions of the Langley Spin Tunnel tests. During spin attempts, airspeed rapidly increased, and stabilized spins could not be obtained. As a result of this spin resistant behavior, the T-45 is not used for spin training. (The T-2 and TA-4 had been used for spin training.)

Ground Handling

In 1999, the Navy requested that Langley provide data, analysis, and recommendations to eliminate a severe pilot-induced oscillation (PIO) resulting in erratic yaw behavior of the T-45 during runway rollout in land-based landing operations. The Navy also requested assistance in determining the cause of uncontrollable aircraft yaw when tires failed.

In response to the Navy requests, Langley researcher Robert H. Daugherty led a Langley effort that conducted over 160 aircraft main and nose tire tests to define cornering and braking behavior in the T-45. The Langley team conducted on-site tests at Patuxent River with a special truck apparatus in a program that included evaluations of various tire states, including normal and blown tires. The tests determined the effect of aircraft roll attitude on uncommanded tire cornering force, identified the drag behavior of failed main gear tires and the reason for lack of yaw control, and provided models of

The T-45 tire characteristics during landing rollout were studied by Langley.

physical tire behavior for analysis. A particularly valuable output of this effort was a dramatic improvement in the modeling capability of T-45 landing operations and the inclusion of the data generated by Langley in the T-45 rollout and training simulators. After the Langley data had been incorporated in the simulator, the landing dynamics of the simulated aircraft were significantly more realistic. The T-45 Program was especially complementary in recognizing this effort as a timely, critical, and valuable NASA contribution to the program.

Inlet Performance

Flight-test experience with the T-45 has demonstrated that the aircraft sometimes experiences undesirable propulsion system characteristics during certain maneuvers. In particular, the aircraft engine has experienced self-clearing "pop" stalls, pop surges, and occasional locked-in surges during simulated air-combat maneuvers and recovery maneuvers from aircraft (wing) stalls.

As a result of concern over these characteristics, the Navy requested that Langley provide support to, and participate in, wind-tunnel tests of the engine inlet behavior of the T-45. In support of this request, a Langley, Navy, and Boeing Team conducted two separate T-45 test entries (May 1998 and July 1999) in the Langley 16-Foot Transonic Tunnel. Langley researcher E. Ann Bare led the test operations.

The 0.26-scale inlet distortion model used in the tests consisted of the T-45 forebody, fuselage contour to the wing trailing edge, and wings. The model was powered with an aft-mounted ejector and included extensive instrumentation to measure steady state and dynamic pressure distortion characteristics. The scope of the test program included evaluations of minor and moderate inlet changes for the cruise and power approach configurations of the T-45.

T-45 inlet performance model in the Langley 16-Foot Transonic Tunnel.

Fairchild Republic A-10 Thunderbolt II

SPECIFICATIONS

Manufacturer
   Fairchild Republic
Date in service
   March 1976
Type
   Close support and attack
Crew
   One
Engine
   General Electric TF34-GE-100A

USER

U.S. Air Force

DIMENSIONS

Wingspan . . . . 57.5 ft
Length . . . . 53.3 ft
Height . . . . 14.7 ft
Wing area . . . . .506 sq ft

WEIGHT

Empty . . . . .18,591.0 lb
Typical combat . . . .23,498.0 lb

PERFORMANCE

Max speed . . . . . Mach number
                                of 0.56
Radius of action . . . . . 695 n mi

HIGHLIGHTS OF RESEARCH BY LANGLEY FOR THE A-10

1. At the request of industry and the Department of Defense, Langley personnel participated in reviews and assessments during competitive stages of Attack Experimental (A-X) Program.

2. Langley conducted wind-tunnel tests and provided data for all competing designs in A-X Program.

3. Langley and Fairchild conducted joint tests of a powered A-10 model in the 7- by 10-Foot High-Speed Tunnel to determine the effects of engine exhaust on aerodynamic characteristics.

4. Fairchild consulted with Langley personnel on A-10 wing airfoils, stall characteristics, high-lift aerodynamics for low-level maneuvers, and spin characteristics.

5. Langley conducted extensive spin tunnel tests of the A-10 for a variety of external store configurations.

The Fairchild Republic A-10 was the result of the Air Force's A-X competition that began in 1967 for a close-support aircraft capable of withstanding the highly lethal environment near the front lines. During the A-X competition, the Department of Defense (DOD) requested that Langley participate in mission analyses, conduct supporting wind-tunnel tests, and provide analyses of the competing designs. During the summer of 1967, tests were conducted in the Langley 7- by 10-Foot High-Speed Tunnel on designs submitted by Grumman, Northrop, McDonnell, and General Dynamics. In 1970, the requirements for the A-X mission were changed, and the Air Force issued a new request for proposals (RFP). Six companies responded to the RFP, and designs by Fairchild and Northrop were selected for further development. In preparation for a competitive fly off under the "fly before buy" philosophy, Langley conducted performance and spin tunnel tests of the Fairchild Republic and Northrop configurations, known as the YA-10A and the YA-9A, respectively.

After the Air Force selected the A-10 as its close-support aircraft in January 1973, Langley conducted extensive tests in the Langley 20-Foot Vertical Spin Tunnel of the production configuration with a large variety of external stores. Langley facilities used in support of the A-10 Program also included the 7- by 10-Foot High-Speed Tunnel.

LANGLEY CONTRIBUTIONS TO THE A-10

The A-X Competition

In the Vietnam conflict concentrated small-arms fire, ground-to-air missiles, and other more sophisticated defenses were particularly lethal to aircraft flying close-support missions. This situation resulted in dramatic changes in philosophy for the capabilities of aircraft conducting these missions. Specialists at Langley were asked to identify technologies that could be applied to a hard-maneuvering aircraft with a relatively high degree of survivability. In March 1967, the Air Force released a request for proposals (RFP) for a new Attack Experimental (A-X) Program. In 1967, numerous wind-tunnel tests were conducted in the Langley 7- by 10-Foot High-Speed Tunnel to obtain analysis data for designs by Grumman, Northrop, McDonnell, and General Dynamics. The tests were led by William P. Henderson and Linwood W. McKinney, who conducted extensive briefings for the Department of Defense (DOD) and the industry competitors on the merits of each configuration.

In the years following 1967, the A-X mission requirements began to change as the threat of Soviet armor and all-weather operations became embedded in military priorities. In 1970, a new A-X RFP was released by the Air Force and six companies responded. After a formal study, Northrop and Fairchild were named as final competitors for a fly-off competition at Edwards Air Force Base.

In 1971, tests of a powered model of the Fairchild A-10 were conducted in the 7- by 10-Foot High-Speed Tunnel by Vernon E. Lockwood. The principal objective of the tests was to determine the effects of engine exhaust flow on aerodynamic characteristics in high-power conditions. The tests were conducted with a special model installation in which the powered engine nacelles were not mounted directly to the airframe, but to a separate sting that included air supply lines for the simulation of power. Critical data were obtained on the impact of power on performance and stability of the configuration, including ground effects.

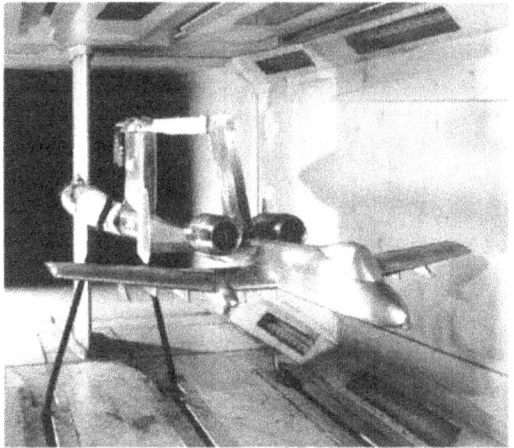

Powered-model tests of the Fairchild A-10 with engines supported by an auxiliary upper sting in the Langley 7- by 10-Foot High-Speed Tunnel.

Spin Recovery

In 1972, tests were conducted in the Spin Tunnel by Stanley H. Scher for the competing Northrop (A-9) and Fairchild (A-10) A-X configurations. Results of these tests were used for evaluations and planning for the competitive fly off that occurred at Edwards Air Force Base later that year. In addition to providing information on advantages and disadvantages of each configuration, these results provided spin and recovery characteristics for pilots of the respective vehicles.

The Fairchild aircraft was announced as the winner of the competition on January 18, 1973, and a formal request for tests in the Langley Spin Tunnel for the development program was received from the Air Force. A large number of tests were conducted because of the variety of external stores, tanks, and armament carried by the aircraft.

High Reynolds number tests were performed with an A-10 model at spin attitudes in the NASA Ames Research Center 12-Foot Pressure Tunnel prior to the spin tunnel tests to determine whether the model would exhibit significant scale effects, and if so, what modifications might be made to correct for these effects. Results of the tests at Ames showed no appreciable scale effects for angles of attack below about 75 deg, but appreciable scale effects (more prospin yawing moments at low Reynolds numbers) for angles of attack above 75 deg. Nose strakes were sized to artificially compensate for the scale effects, but no flat spins (angle of attack greater than 75 deg) were evident and the strakes were not used.

Spin tunnel tests of the A-10 configuration.

General Dynamics F-111 Aardvark

SPECIFICATIONS

Manufacturer
General Dynamics
Date in service
June 1967
Type
Fighter-bomber
Crew
Two
Engine
F-111F . . . Pratt & Whitney
TF30-P-100

USERS

U.S. Air Force (retired 1996)
and Australian Air Force

DIMENSIONS

Wingspan
unswept. . . . . 63.0 ft
fully swept . . . . 31.9 ft
Length . . . . 73.5 ft
Height . . . . 17.1 ft
Wing area . . . . 525 sq ft

WEIGHT

Empty . . . . 47,481 lb
Gross . . . . 100,000 lb

PERFORMANCE

Max speed . . . . . . above Mach
number of 2
Range. . . . . 2,925 n mi

HIGHLIGHTS OF RESEARCH BY LANGLEY FOR THE F-111

1. Langley conducted studies to enable the variable-sweep wing that was employed by the F-111 and subsequent variable-sweep aircraft such as the F-14 and B-1.
2. Langley staff conducted wind-tunnel tests, identified major problems, and recommended critical improvements to the aerodynamic performance of the F-111.
3. Langley personnel presented key technical results in congressional and high-level Department of Defense investigations of the technical merit of the F-111.
4. Langley led studies to provide solutions to major propulsion integration problems exhibited by the F-111, which were used on all subsequent high-performance fighters, such as the F-14, F-15, and F-22.
5. Langley identified critical deficiencies in the high-angle-of-attack characteristics of the F-111.
6. Langley detected the presence of a dangerous flat spin for the F-111 and precipitated the development of a stall inhibitor system to protect against inadvertent spins.
7. Following a catastrophic wing fatigue failure of an operational F-111, Langley conducted critical materials tests, and Langley personnel served on advisory boards to recommend approaches to ensure safety of the fleet.
8. Langley conducted tests to improve the robustness and landing loads of the F-111 crew module.
9. Langley wind-tunnel tests to determine the flutter characteristics of the F-111 identified a potential tail flutter problem that resulted in modifications to the aircraft.
10. Langley supercritical wing technology was applied to the F-111 and demonstrated during the TACT and MAW Programs.

The General Dynamics F-111 was a multipurpose supersonic tactical fighter-bomber aircraft. The F-111 was one of the more controversial aircraft in the U.S. military inventory, yet it achieved one of the safest operational records in Air Force history. It was regarded as a highly effective, all-weather interdiction aircraft. As the result of a high level government directive in 1961, both the Navy and Air Force became committed to a common Tactical Fighter Experimental (TFX) Program. The TFX Program called for developing a single aircraft to fulfill a Navy fleet defense interceptor requirement and an Air Force supersonic strike aircraft requirement. The mission requirements were impossible to achieve, especially since planners placed priority upon the Air Force requirement, and then tried to tailor a heavy land-based aircraft to the demands of carrier-based naval aircraft. The naval version, the F-111B, was never placed into production. The Air Force aircraft was produced in a variety of models, including the F-111A, F-111D, F-111E, and F-111F fighter-bombers; the FB-111A strategic bomber; the F-111C for the Australian Air Force; and an EF-111 electronic warfare version. The U.S. Air Force versions were retired in 1996, but the Australians plan to operate their fleet until well into the twenty-first century.

Arguably, the political and technical issues associated with the F-111 program resulted in more research activities at the Langley Research Center than any other production military aircraft in Langley's history. The staff participated in top level assessments of the aircraft's capabilities (including briefings at the highest levels of the Department of Defense (DOD) and government), identified several critical problems, and provided recommendations for solutions. Significant Langley technical contributions included the variable-sweep wing concept, and test activities in aerodynamic performance enhancements, high-angle-of-attack characteristics, spin recovery, flutter, propulsion integration, wing structures and materials, and robustness of the crew escape module. Finally, Langley was a major participant in the joint NASA and Air Force F-111 Transonic Aircraft Technology (TACT) Program and the Mission Adaptive Wing (MAW) Program that applied supercritical wing technology, which was developed at Langley, through flight tests of a modified F-111 research aircraft at NASA Dryden Research Center.

Langley facilities that contributed to the F-111 program included the 7- by 10-Foot High-Speed Tunnel, the 4- by 4-Foot Supersonic Pressure Tunnel, the Unitary Plan Wind Tunnel, the 16-Foot Transonic Tunnel, the 8-Foot Transonic Pressure Tunnel, the 26-Inch Transonic Blowdown Tunnel, the 30- by 60-Foot (Full-Scale) Tunnel, the 20-Foot Vertical Spin Tunnel, the 16-Foot Transonic Dynamics Tunnel, the Fatigue and Fracture Laboratory, the Impact Dynamics Research Facility, the 12-Foot Low-Speed Tunnel, and radio-controlled drop models.

LANGLEY CONTRIBUTIONS TO THE F-111

The Variable-Sweep Wing Concept

The F-111 was the first production military aircraft to capitalize on years of Langley research to develop a variable-sweep wing aircraft. Variable-sweep wings provide significant benefits to the aerodynamic performance of aircraft that operate over a wide range of altitudes and airspeeds. An excellent summary of the history of Langley's role in the development of this breakthrough concept was given by Edward C. Polhamus (contributor to variable-sweep research and the F-111 program) in his 1983 American Institute of Aeronautics and Astronautics (AIAA) Wright Brothers Lecture (ref. 4).

As the advantages of wing sweep for enhanced aerodynamic performance became known near the end of World War II, designers considered several approaches to providing variable sweep for efficient characteristics at both low and high speeds. The German Messerschmitt P-1101 research configuration had provisions for three ground-adjustable sweep angles; however, the aircraft was only carried to the prototype stage and flight tests were never conducted. Although it never flew, the P-1101 was captured and brought to the United States, where it was later evaluated by Bell Aircraft Corporation in a study that strongly influenced the design of the Bell X-5 research aircraft.

The first wind-tunnel tests of variable-sweep concepts were conducted at Langley in the mid-1940's. These tests included the now familiar symmetric variable-sweep wing, as well as the variable oblique-wing concept (free-flight model tests by John P. Campbell). Researcher Charles J. Donlan, who would later become Deputy Director of Langley, initiated tests of an existing model of the famous Bell X-1 (first aircraft to exceed the sound barrier) in the Langley 300-MPH 7- by 10-Foot Tunnel in 1947 to explore the challenges of variable sweep. The results of the study, which used a single centerline pivot for the wing, clearly showed that such an arrangement resulted in excessive longitudinal stability and marginal maneuverability with the wings swept aft. To be successful, the variable-sweep concept would have to minimize the dramatic increase in stability. Several methods were explored, and it was concluded that some type of variable longitudinal translation of the pivot point was required—although this solution was undesirable from a weight and complexity perspective.

Time-lapse photograph of the Bell X-5 shows the wing at various sweep angles.

The radical British Swallow configuration in the Langley 16-Foot Transonic Tunnel with variable sweep arrow wing and over and under wing-mounted engines.

In the late 1940's, under somewhat reluctant Air Force sponsorship and based on the German and U.S. studies, the Bell X-5 became the Air Force and the National Advisory Committee for Aeronautics (NACA) workhorse for variable-sweep studies. Wing sweep on the X-5 could be continuously varied from 20 deg to 60 deg through the use of a translating wing-pivot mechanism. The X-5 configuration was tested in the Langley 300-MPH 7- by 10-Foot Tunnel, the 8-Foot Transonic Pressure Tunnel, the 4- by 4-Foot Supersonic Pressure Tunnel, and the 20-Foot Vertical Spin Tunnel. The first flight of the X-5 was made on June 20, 1951. At about the same time, interest in variable sweep was growing in the United Kingdom, and discussions between Langley Assistant Director John Stack and the British resulted in exchanges of data and proposed cooperation for research efforts.

In 1952, the Grumman XF-10F variable-sweep aircraft began flight tests. (The XF-10F had also been tested in the 7- by 10-Foot High-Speed Tunnel at Langley.) Although the aircraft was underpowered and subsonic, it demonstrated the aerodynamic performance advantages of variable sweep. However, the wing-pivot translation feature added considerable weight to the configuration, and the performance of the XF-10F could not compete with the rapid increase in supersonic performance that was being demonstrated on most military aircraft of its size.

At this point, military interest in the variable-sweep concept rapidly decreased. It appeared that supersonic flight could not be sustained with state of the art engines at that time. Also the heavy, complicated translating wing-pivot mechanism was not an acceptable penalty when compared with the performance of a moderately swept fixed wing. However, as has been the case throughout Langley's history, a few visionaries correctly anticipated the future military aircraft requirements for sustained supersonic flight and successfully advocated for fundamental research to continue to develop and optimize the variable-sweep concept. Led by John Stack and Charles Donlan, Langley researchers joined with the British in a North Atlantic Treaty Organization (NATO) sponsored cooperative study of variable-sweep concepts that included extensive tests in the Langley 7- by 10-Foot High-Speed Tunnel and the 16-Foot Transonic Tunnel.

The British and Langley cooperative tests in 1958 included work on the British Swallow configuration, which was a radical tailless slender arrow-wing configuration with variable-sweep wings and pivoting wing-mounted engines. The Swallow exhibited numerous stability and control problems. Langley brought three variable-sweep configurations to the study, including two that used pivot locations in the fuselage near the trailing edge of the inner wing and folding tail control surfaces to maintain stability levels. Thomas A. Toll, responsible for the stability portion of the variable-sweep program, assigned William J. Alford, Jr. and Edward Polhamus to the task. In 1959 they arrived at the breakthrough solution, which was to locate the pivots of the movable wing panels to positions outboard of the fuselage. With this outboard wing-pivot arrangement, sharing of lift between the fixed inner wing and the movable outer wing panels minimized the movement of the aerodynamic center of lift. A fourth configuration, known as configuration IV, was added to the research program to confirm this design concept. The configuration was tested across the speed range in Langley tunnels with great success.

Encouraged by their success, the Langley team conducted in-depth analyses of variable-sweep aircraft representative of configurations for Navy and Air Force missions. Tests of all types—force and moment, dynamic, and free flight—were conducted in virtually all of the major tunnels at Langley. Variable-sweep configurations applicable to the Navy combat air patrol mission were studied in a program known as CAP, and the Air Force tactical aircraft mission was the basis for configurations studied in a program known as TAC. Extensive wind-tunnel tests and analysis by Langley researchers had matured the variable-sweep concept.

Alford and Polhamus became internationally recognized for their research on the outboard wing-pivot concept, and they mutually hold the U.S. patent on this revolutionary discovery, which has been successfully applied to numerous U.S. and foreign military aircraft.

Configuration IV, which was the breakthrough model for variable-sweep technology.

James L. Hassell, Jr. with a free-flight CAP model that was flown in the Langley Full-Scale Tunnel.

*The Tactical Fighter
Experimental (TFX) Program*

In 1957, the U.S. Navy requested industry responses for the design of a low-altitude strike fighter. John Stack briefed senior Navy managers that a proposed British low-altitude strike fighter, the NA-39, would be much more advanced than the Navy aircraft. He also suggested the application of variable sweep to leapfrog the capabilities of the NA-39. Following briefings by Langley personnel to the Navy, the mission specifications for the new Navy fighter were expanded to include multimission capability with a requirement that variable-sweep applications be studied. The request for proposals went to industry in early December 1959 and set the stage for what would ultimately become the Tactical Fighter Experimental (TFX) Program.

Meanwhile, the Air Force Tactical Air Command (TAC) Requirements Division at Langley Air Force Base (adjacent to the Langley Research Center) was attempting to define a replacement for the F-105 fighter-bomber aircraft. TAC was interested in an aircraft that could carry nuclear weapons internally, fly transatlantic routes without refueling, operate from semiprepared fields in Europe, have a top speed of Mach number of 2.5 at high altitudes, and fly at high subsonic speeds at low altitudes. The aircraft would perform a "low-low-high" mission, wherein it would cruise into the vicinity of the target at low altitudes and subsonic speeds, perform a low-altitude dash to the target at high subsonic speeds, and perform a high-altitude, long-range cruise back to base at subsonic speeds. The Mach number of 2.5 capability would be used for high-altitude engagements against enemy fighters. Initial analysis by industry of the request indicated that a fixed-sweep aircraft capable of meeting the requirements would weigh in excess of 100,000 lb (too heavy for unprepared fields) and demand the attributes of low sweep for transatlantic flight, but high sweep for the high-speed requirements. TAC was therefore in a stalemate without a viable design approach to its requirements.

John Stack approached the TAC planners in 1959 with the benefits of variable sweep to enable an aircraft to meet the requirements. The extended ferry range that is provided by variable sweep was of prime importance to TAC, since estimates indicated that transatlantic range might be possible. Together with the commander of TAC, Stack laid out a realistic set of aircraft performance requirements that included the desired low-altitude dash capability at high subsonic speeds. Unfortunately, as the requirements went through the TAC system for approval, the final specifications called for a 210-n-mi, sea level dash at a speed that had increased from a Mach number of 0.9 to a Mach number of 1.2. Upon learning of the supersonic low-altitude speed requirement, Langley quickly informed the Air Force that this capability was impossible to meet for the range specified. Nonetheless, TAC was committed to the unrealistic specification. (In flight tests of the F-111A in 1969, the actual low-altitude supersonic dash performance of the aircraft was only 30 n mi.)

In 1960, the Air Force and the Navy were both attempting to develop new fighter aircraft. The Kennedy administration's Secretary of Defense, Robert McNamara, ordered the development of a single aircraft for both the Air Force and the Navy (to be led by the Air Force), called the Tactical Fighter Experimental (TFX). Mr. McNamara defined the basic mission requirements when the Air Force and Navy could not agree, and in October 1961, a request for proposals (RFP) was issued to industry. Boeing won all four stages of the competition that followed, but McNamara overruled the source selection board and decreed on November 24, 1962, that the General Dynamics and Grumman Team would build the TFX.

Langley's variable-sweep wing concept was nationally recognized as the technical key that would unlock the capabilities of the TFX. McNamara said (ref. 27)

New developments in engine performance and in aerodynamics, particularly the variable-geometry wing concept evolved by NASA, now make it possible to develop a tactical fighter that can operate from aircraft carriers as well as from much shorter and cruder runways, and yet can carry the heavy conventional ordnance loads needed in limited war.

As an example of the plaudits given Langley for the variable-sweep contribution to the evolving TFX program, the following comments of editor Robert Hotz of the internationally acclaimed *Aviation Week* magazine appeared in the magazine on December 3, 1962:

Underlying the whole TFX concept is one of the solid, basic technical explorations of the old National Advisory Committee for Aeronautics (NACA) that did so much to keep this country the international leader in supersonic aircraft development. Without the fundamental research into the variable-sweep wing and the detailed development of this principle by the Langley research laboratory group headed by John Stack, the current TFX concepts of both final competitors would have been impossible... When Congress convenes again and begins carping over the Fiscal 1964 NASA budget for aeronautical research, the full story of the Langley contributions to the TFX program should be hammered home as an example of how these research and development investments eventually pay substantial benefits.

After the TFX contract was awarded, Langley, Ames, Glenn, and Dryden all supported the F-111 development program. Because of the strong interest in this aircraft, and the large magnitude of NASA support, the program was rigorously managed and documented, beginning in November 1962. Mark R. Nichols, Laurence K. Loftin, Jr., Edward Polhamus, Jack F. Runckel, Theodore G. Ayers, and M. Leroy Spearman were the leaders and spokesmen for the F-111 support activities at Langley. In addition, Polhamus served as the Langley focal point for overall F-111 activities and spent considerable personal time in coordination of tunnel requests and NASA, industry, and DOD joint meetings.

In 1963, political turmoil surfaced as a special Senate subcommittee chaired by Senator McClellan of Arkansas held hearings on the award of the TFX Program. This committee gathered a considerable amount of Langley data, briefings, and testimony during various phases of the hearings, which took place over several years.

For the first 3 years of the F-111 development program (1963 to 1965), a total of nearly 20,000 hr of tests were conducted in NASA wind tunnels (about 15,000 hr at Langley). Over 15 wind tunnels were utilized, making the F-111 program the most extensive wind-tunnel support effort ever provided for one aircraft by NASA or the NACA. By 1968, over 22,000 hr of tunnel tests had taken place at Langley. (In contrast, Langley expended 5,000 hr for development of the F-105.) The large number of test hours was the result of the multiple versions of the aircraft (F-111A, F-111B, RF-111, and FB-111), the addition of wing sweep as a test variable, a vast number of external store configurations, and concentrated technical assaults on a multitude of problems—especially the transonic drag issue. During the development effort, the Langley staff also participated in numerous F-111 advisory and assessment teams and briefings for DOD, Congress, and industry. For example, Edward Polhamus presented the scope and results of the NASA effort to the McClellan Committee during its second session in 1970.

The first F-111A flew in December 1964, and the first F-111B flew in May 1965. The most positive result from early flight evaluations was the very satisfactory behavior of the variable-sweep wing system. However, the aircraft were judged to be sluggish and underpowered. Furthermore, the engines exhibited violent stalling and surging characteristics.

An outstanding, in-depth discussion of the details of the initiation and early years of the F-111 development program is given in the book *Illusions of Choice* by Robert F. Coulam and Robert S. McNamara (ref. 27).

Aerodynamic Performance

On December 19, 1962, representatives of General Dynamics and Grumman visited Langley for discussions of the supersonic performance of the F-111. The manufacturers were informed that the supersonic trim drag of the aircraft could be significantly reduced and maneuverability increased by selecting a more favorable outboard wing-pivot location. Unfortunately, the manufacturers did not act on this recommendation, and it was subsequently widely recognized that the F-111 wing pivots were too far inboard. (It should be noted that the F-14 designers, aware of this shortcoming, designed the F-14 with a more outboard pivot location.) The F-111 subsequently exhibited very high levels of trim drag at supersonic speeds during its operational lifetime.

In March 1963, the initial supersonic tests of the F-111 in the Langley Unitary Plan Wind Tunnel by David S. Shaw confirmed the Langley expectations of high trim drag at supersonic speeds. A month later, tests conducted by Theodore Ayers of a 1/24-scale F-111 model at transonic speeds in the Langley 8-Foot Transonic Pressure Tunnel indicated that the transonic drag was considerably higher than General Dynamics predictions. Therefore, a large drag reduction had to be accomplished to meet mission requirements. Several discussions between Langley and General Dynamics were also held to define approaches to improve supersonic maneuverability. Langley continued to emphasize the importance of wing-pivot location and recommended a change in pivot location and a forward shift of the wing as a solution to the problem. It was decided that the modified wing suggested by Langley would be built and tested. Supersonic tests of the Langley wing modification indicated a large increase in maneuverability that would allow the F-111 to approach the proposed maneuverability levels. However, because of changes in maximum cross-sectional area, some transonic drag penalty would be

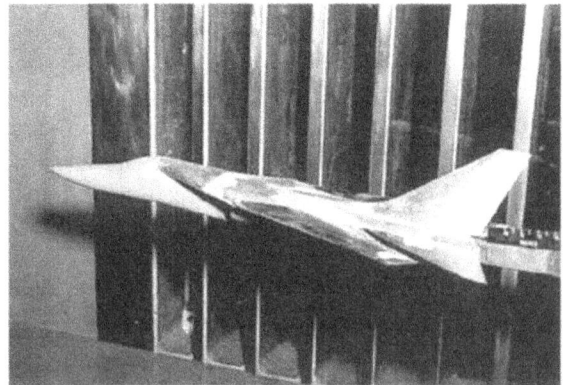

*Supersonic tests of the early F-111 design in the
Langley Unitary Plan Wind Tunnel in 1963*

expected with the modification. At a later meeting in June between Langley, the F-111 Systems Program Office (SPO), and General Dynamics, the positive results of the modified wing were discussed, but the F-111 SPO expressed concerns over any possible transonic drag penalty, the engineering effort required for the change, and potential schedule slippage. The Air Force cancelled further studies of the wing modification.

Meanwhile, a controversy erupted over the discrepancy of transonic drag estimates between the data obtained by Theodore Ayers in the Langley 8-Foot Transonic Pressure Tunnel and data obtained in the Cornell Aeronautical Laboratory 8-Foot Transonic Pressure Tunnel. The Langley staff investigated these differences and concluded that the lower drag measured in the Cornell facility was probably due to interference effects caused by an oversized model and the large, blunt support sting. Langley offered to investigate the problem by testing a mock-up of the Cornell support system. In October 1963, special tests were conducted in the Langley 8-Foot Transonic Pressure Tunnel with a mock-up of the Cornell support system. The results of the test indicated a very large buoyancy effect, which accounted for the erroneous low transonic drag measurements in the Cornell tunnel. As Langley's support for the F-111 continued, the staffs of the 16-Foot Transonic Tunnel and the 8-Foot Transonic Pressure Tunnel remained sensitive to wind-tunnel accuracies, and numerous tests were conducted with the same model in these two tunnels to establish confidence in the Langley projections of transonic drag.

With a much higher level of drag established, emphasis was placed on reducing the transonic drag. The drag problem included large afterbody and nozzle drag components caused by high closure slopes (rapid variations in external aircraft contours), high drag components of the cockpit and inlets, and very high boundary-layer spillage drag. Dr. Richard T. Whitcomb and his staff conducted exhaustive tests of engine alignment, wing and tail twist, and even antishock bodies to reduce drag. Studies by Ayers concluded that the area ruling for the F-111 was unsatisfactory.

Also in 1963, an F-111 aerodynamic consulting group consisting of Air Force, Navy, and NASA members met and concluded that the transonic aerodynamic performance of the F-111 would be considerably below the requirements for the projected missions. Calculations based on Ayers' drag measurements predicted that the aircraft would only have a range of about 20 to 30 n mi for flight at low altitudes and speeds of a Mach number of 1.2, in contrast to the 220 n mi capability predicted by General Dynamics. Actual flight tests later verified Ayers' projection. The group, which included Polhamus and Spearman, recommended that aft-end modifications suggested by Langley should be studied for transonic drag reduction.

As concern over the aerodynamic performance of the F-111 increased, Charles Donlan and Edward Polhamus briefed the Assistant Secretary of the Air Force in April 1964 on the situation. They recommended that the staff of the Langley 16-Foot Transonic Tunnel define the benefits of Langley conceived aft-end modifications. It was also suggested that the wing with the longer span of the Navy aircraft be used on the Air Force aircraft. Polhamus and Spearman also briefed Air Force General Schreiver and Navy Admiral Schoech a month later on the transonic drag problem.

During May 1965, representatives of Grumman visited Langley several times to discuss methods of improving the acceleration and maneuverability of the Navy F-111B. Modifications considered by Grumman included several of the early Langley suggestions, such as a modified wing and pivot location, a straightened tailpipe, and an improved interengine fairing. In addition, Grumman examined a modified horizontal tail, alternate missile arrangements, and an aft-fuselage modification. Although these modifications never came to fruition for the F-111B, the discussions had a large impact on the later design of the F-14 by Grumman, which became an outstanding Navy aircraft.

Langley also assessed the effect of the bomb bay cavity and doors on directional stability in the Unitary Plan Wind Tunnel; the impact of missile carriage at supersonic speeds in the Unitary Plan Wind Tunnel; the aerodynamic damping during subsonic, transonic, and supersonic flight in the 30- by 60-Foot (Full-Scale) Tunnel, the Unitary Plan Wind Tunnel, and the 8-Foot Transonic Pressure Tunnel; the transonic and supersonic flight of the strategic bomber (FB-111) and reconnaissance (RF-111) versions of the F-111 in these same wind tunnels; and the development of the F-111B by the Navy, until it was cancelled on July 10, 1968. Over 25 test entries in the Unitary Plan Wind Tunnel were made for the F-111 variants.

Propulsion Integration

Hot-jet tests of a 1/9-scale ejector nozzle at transonic speeds were conducted in April 1963 in the Langley 16-Foot Transonic Tunnel to begin what would become an extensive series of F-111 propulsion integration studies. (The staff ultimately conducted 17 entries of F-111 models or components during the program.) Later that year, a 1/6-scale inlet model was tested to determine the effects of aircraft nose shape (Air Force and Navy) and engine inlet spike configuration. Following this test entry, the inlet, cowl, and spike geometry were revised.

In 1964, General Dynamics, in consultation with the 16-Foot Transonic Tunnel staff, completed fabrication of a 1/12-scale model designed to investigate propulsion-airframe integration characteristics. This model represented the most realistic and complex model of a military fighter ever tested in the 16-Foot Transonic Tunnel. The model had multiple strain-gage balances and balance arrangements to independently measure thrust, drag, and thrust minus drag. It also contained three independently controlled internal flows to simulate the F-111 blow-in-door ejector exhaust system: a hot hydrogen peroxide primary-jet flow system, a high-pressure air secondary-flow system, and a low-pressure air boundary-layer bleed system. In addition, the model contained fully variable, aerodynamically actuated blow-in doors and ejector shroud flaps on the nozzles. Tests in 1964 on this model in the 16-Foot Transonic Tunnel indicated a significant nozzle-thrust deficiency that was associated with an adverse fuselage afterbody flow field. At the end of the year, additional tests of the hot-jet model were directed at a wide range of ejector nozzle geometries in an attempt to solve nozzle-thrust and flutter problems. Langley also initiated studies directed toward reducing the large base drag associated with the short interengine fairing (interfairing). Relatively large drag improvements were obtained with a long interfairing design conceived by Langley.

Unfortunately, the naval F-111B configuration was too long to met the requirements for aircraft carrier elevator spotting (compatibility of the aircraft dimensions with the elevator on the aircraft carrier that transports aircraft to and from the flight deck and the lower hangar area). Follow-on tests of the 1/6-scale inlet model during 1964 included studies of the effects of an extended nose (for the RF-111), a weapons bay pod, and bleed doors.

Jack Runckel and Edward Polhamus briefed Air Force management in October 1964 on the ejector nozzle problems of the F-111. Since the nozzle problem was associated with the aircraft aft-end flow field and since the aft-end drag was relatively high, the Langley representatives recommended that improvements to the back end of the aircraft be investigated as a high priority item. Runckel subsequently became a key figure in a joint F-111 nozzle committee that included the F-111 SPO, the Bureau of Naval Weapons, Pratt and Whitney, General Dynamics, and NASA. In a meeting of this committee in

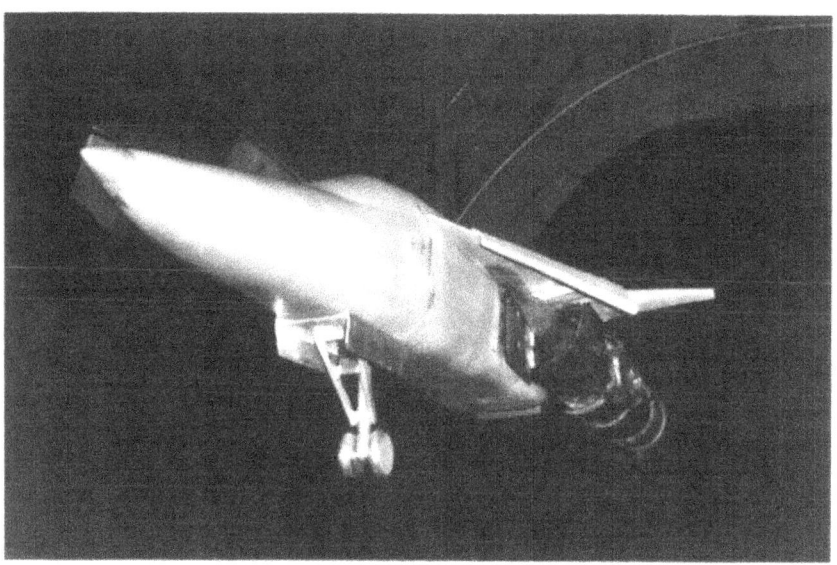

Inlet tests of an F-111 model in the Langley 16-Foot Transonic Tunnel.

January 1965, Runckel proposed that a truncated, concave base interfairing be investigated as a means of reducing drag at transonic speeds, while still meeting the length restriction imposed by the F-111B naval version.

In February 1965, tests of the 1/12-scale hot-jet model indicated large improvements in transonic performance because of airframe changes in the region between the nozzles. This model configuration had the best aerodynamic characteristics to date in the program. A few months later, a meeting was held at Langley with representatives from the F-111 SPO, the Bureau of Naval Weapons, General Dynamics, Grumman, Pratt and Whitney, and Langley. At that meeting, it was agreed that emphasis should quickly move to reducing transonic drag by developing an optimum engine interfairing and speed bumps (additional area added to shape the aircraft to comply with Whitcomb's area rule). In December 1965, the hot-jet model was tested to develop the interfairing and define a configuration that would be flight tested in early 1966. The aft-end modifications from this effort were ultimately adopted and resulted in a significant improvement in transonic drag.

The early F-111A exhibited numerous engine problems, including compressor surge and stalls. NASA was a participant in finding solutions to these problems, as its pilots and engineers flew test flights of the aircraft to determine inlet pressure fluctuations (dynamics) that led to these events. Eventually, as a result of NASA, Air Force, and General Dynamics studies, the engine problems were solved by a major inlet redesign.

The F-111 program brought many difficult challenges in propulsion-airframe integration to the staff of the Langley 16-Foot Transonic Tunnel. These problems were extremely complex and demanded timely solutions in a highly visible, controversial national program. However, the staff responded with outstanding technical expertise and innovation. By the end of 1967, 19 F-111 related model entries had been completed in the 16-Foot Transonic Tunnel, and over 283 configurations were investigated during 12.5 months of tunnel occupancy. This effort ultimately required participation of almost every 16-Foot Transonic Tunnel staff member. Staff members who made significant contributions to the F-111 program included Richard J. Re, Odis C. Pendergraft, Jr., Richard G. Wilmoth, Charles E. Mercer, Francis J. Capone, and Bobby L. Berrier. As a result of the F-111 program, the research capability on aft-end transonic drag problems greatly increased. This research capability contributed to all subsequent high-performance military aircraft and placed Langley in a position of world leadership in this critical technical area.

High-Angle-of-Attack
Characteristics

Initial free-flight model tests of the F-111 configuration were led by Peter C. Boisseau in the Langley 30- by 60-Foot (Full-Scale) Tunnel in October 1964. During the flight tests, the wing sweep was varied from 16 deg to 72.5 deg. In December, the model was flown to determine the effects of stability augmentation in roll and pitch for the clean and landing configurations. The flight tests were extended to high angles of attack, including the stall.

When the model was flown to high angles of attack with the wings at the 50-deg and 72.5-deg sweep conditions, the model exhibited a sudden, uncontrollable yaw divergence prior to maximum lift. General Dynamics personnel, including the test pilot who was scheduled to make the first high-angle-of-attack flights with the aircraft, witnessed the tests.

The model free-flight tests also indicated an unusual unsteadiness in lateral behavior at moderate angles of attack for the landing configuration. The unsteadiness was apparently caused by an unsteady flow off the wing root glove (the fixed, highly swept inner wing). In early 1965, extensive flow visualization tests were made in the Langley 12-Foot Low-Speed Tunnel in an effort to change the vortex-flow field set up by the glove and to delay separation on the inner wing for the landing configuration. This work

Time-lapse photograph of the F-111A free-flight model at several
wing sweep angles during tests in the Langley Full-Scale Tunnel.

included an investigation of a rotating glove vane, which was then evaluated during the free-flight model test in the Full-Scale Tunnel. The glove vane cured the roll unsteadiness previously noted and was subsequently incorporated into the F-111 landing configuration.

As F-111 operations expanded within the Air Force in the late 1960's, a rash of incidents involving unexpected departures from controlled flight during maneuvers at high angles of attack occurred. The Air Force requested industry and NASA assistance in analyzing and solving the problem, which was viewed as a significant flight safety issue. Langley researchers Joseph R. Chambers and James S. Bowman, Jr. served on an Air Force, industry, and NASA committee that identified a shortcoming in the F-111 flight control system that promoted the unintentional departures. The F-111 had been designed with a g-command flight control system that provided g-forces in direct proportion to the deflection of the pilot control stick. However, in providing the pilot with the level of g-force, the system would increase the angle of attack of the aircraft. Unless the pilot was monitoring the angle of attack, the aircraft could enter a range of high angles of attack where a loss of directional stability resulted in an unintentional yaw departure and spin entry. These findings led to an Air Force program in 1973 to develop a stall inhibitor system (SIS) for the F-111. Langley participated in the design and analysis of this system. The SIS was designed to automatically monitor and limit the angle of attack of the aircraft during flight maneuvers and was incorporated into the F-111 fleet.

Spin Recovery

Another area of controversy within the F-111 development program arose in the early 1960's—in the area of spin and spin recovery technology. The conventional approach to assessing and improving spin characteristics of new military aircraft was to conduct tests of dynamically scaled models in the Langley 20-Foot Vertical Spin Tunnel to determine spin and spin recovery characteristics, as well as the size of the emergency parachute required for spin test aircraft. In conjunction with these tests, radio-controlled drop-model tests were conducted to assess spin-entry tendencies and to assess the effectiveness of out-of-control recovery procedures.

For the F-111 program, General Dynamics proposed to assess spin recovery characteristics with analytical methods, in lieu of the traditional dynamic scale model tests. In addition, General Dynamics requested over 1,000 hr in Langley tunnels to provide the data required for the proposed study. Langley's reaction to this proposal was extremely negative because analytical procedures for spin studies had not been validated, and in the opinion of James Bowman and the Spin Tunnel staff, could not be trusted for such an important aircraft program. However, the Air Force accepted Langley's recommendation to terminate the General Dynamics plans for the analytical approach and elected to continue with the traditional Spin Tunnel and drop-model tests.

At the request of Langley, a 1/24-scale model of the F-111 was tested in the Ames 12-Foot Pressure Tunnel in early 1964 to examine Reynolds number effects at high angles of attack and sideslip. The results of these tests necessitated forebody modifications for the spin tunnel model to simulate, at the low Reynolds numbers of spin tunnel investigations, the cross-flow characteristics approximating the Reynolds number conditions of full-scale flight.

During the period from October 1964 to May 1966, extensive spin tunnel tests were made by James Bowman and Louis White on 1/40-scale models of the F-111A and F-111B aircraft to determine spin and recovery characteristics. Tests were conducted for wing-sweep angles of 20 deg, 26 deg, 50 deg, and 72.5 deg. The results of the tunnel tests indicated that the F-111 would exhibit several spin modes, including steep oscillatory spins from which recovery could be accomplished and a fast flat spin from which recovery was marginal or impossible using aerodynamic controls. The flat spin was especially stable. A number of radical approaches to breaking the spin were attempted, including sweeping the wings forward and rearward during the spin. For these tests, the model was equipped with a small electric motor that was remotely actuated to drive the wing-sweep angle. However, the flat spin could not be slowed or stopped using this technique. All of these spins (including the flat spin) were subsequently encountered in F-111 spin tests at Edwards Air Force Base and in fleet operations. Bowman served on several spin accident investigation committees formed by the Air Force.

Extensive studies were made of the spin entry and post-stall motions of the F-111A by Charles E. Libbey with two 1/9-scale helicopter drop models. Over 50 successful drops were made with wing sweeps varying from 16 deg to 72.5 deg. Results of these tests showed that certain pilot inputs following the yaw departure at high angles of attack (as previously discussed for the wind-tunnel free-flight model) would promote the fast flat spin. These results were discussed with Air Force representatives for inclusion in the pilot handbook procedures for avoiding this extremely dangerous condition.

Several F-111 aircraft were lost in spin accidents during fleet operations; however, the subsequent implementation of the SIS prevented stalls and eliminated spins as an operational concern.

During the late 1960's, Langley researcher William P. Gilbert conducted fundamental research on automatic spin prevention systems for fighter aircraft. Gilbert's work was stimulated by the fact that flight control systems were beginning to use flight parameters (such as angle of attack and yaw rate) that would permit the mechanization of spin prevention for routine operations in highly redundant systems. Previously, the concept of

The F-111A drop model prior to launch from a helicopter for a spin-entry test.

automatic spin prevention was a highly desirable concept, but the mechanization would have required a special system that might be prone to failure and would only be utilized on very rare occasions. With the emergence of the new operational control system components, Gilbert became interested in demonstrating the effectiveness of spin prevention systems. Following a series of analytical studies, Gilbert teamed with Charles Libbey to implement and test the prototype system on an F-111 drop model. With the system engaged, the spin-prone model could be maneuvered to extreme angles of attack without entering a spin, even with full prospin control inputs from the pilot.

On one occasion, Gilbert briefed NASA Administrator, Dr. James Fletcher, on the very positive results of his study. Fletcher praised the work as a forerunner of future systems that would enable carefree maneuvering of military aircraft. Gilbert's work with the F-111 model represented one of the first efforts to develop the highly sophisticated control systems that are now used in virtually every domestic and foreign fighter aircraft to prevent spins during strenuous air combat maneuvers.

Flutter Tests

In early 1963, flutter trend models were tested in the Langley 26-Inch Transonic Blowdown Tunnel to determine the general flutter boundary for the isolated wing and tail surfaces of the F-111. Several tests were conducted in this facility, before a dummy (stability) model of the complete F-111 configuration was tested in the Langley 16-Foot Transonic Dynamics Tunnel (TDT) in August 1963. The initial TDT tests were to check out the wind-tunnel suspension system for future flutter tests. Subsequent tests of the isolated wing and horizontal tail in the Blowdown Tunnel revealed that the horizontal-tail design had an inadequate margin of safety for flutter at low supersonic speeds, and the geometry of the F-111 tail configuration was changed.

Flutter clearance tests of the F-111 empennage model and a 1/8-scale complete flutter model of the F-111 were made in the TDT during February 1965. Flutter tests continued through 1965 and subsequent years to examine the effects of a number of external store configurations on flutter boundary. External stores for the F-111 included combinations

F-111 model in the Langley 16-Foot Transonic Dynamics Tunnel in 1963. (Tail surfaces were later changed to avoid flutter.)

*FB-111 flutter model illustrating the wide scope of
external stores that were tested in the program.*

of bombs, missiles, and fuel tanks. The wing pylons pivoted as the wings swept back, keeping the ordnance parallel to the fuselage. F-111 tests led by Charles L. Ruhlin, Maynard Sandford, and Irving Abel ultimately included 13 test entries in the TDT of versions such as the F-111A, F-111B, and FB-111.

Crew Escape Module

The two crew members in the F-111 sat side by side in an air conditioned, pressurized cockpit module that served as an emergency escape vehicle and a survival shelter on land or in water. In emergencies, crew members remained in the cockpit, an explosive cutting cord separated the cockpit module from the aircraft, small rocket engines ejected the module from the aircraft, and the module descended by parachute. The ejected module included a small portion of the inner wing glove to stabilize it during aircraft separation. Air bags cushioned the landing impact and helped to keep the module afloat in water. After separation, the air bags were inflated with nitrogen (stored behind the pilot's seat).

The module could be released at any speed or altitude—even under water. For underwater escape, the air bags raised the module to the surface after it had been severed from the plane. Initial concerns for the module design centered on potential windblast; however, this threat was properly addressed in the design. Unfortunately, impact of the level of acceleration on the crew during landing proved to be a problem. The nominal descent rate of the module on the parachute was about 32 ft/sec.

As a result of a number of back injuries to crew members during use of the recovery system and undesirable postimpact overturning, the Air Force requested that the NASA Crash Dynamics Group conduct drop tests of the F-111 crew escape module at the unique Langley Impact Dynamics Research Facility (IDRF). The facility had initially been used to train astronauts for moon walks as part of the Apollo Program, when it was known as the Lunar Landing Training Facility. After the successful Apollo Program, the Langley staff recognized the value of the tall, large gantry structure for simulating ground impact of large aircraft structures and full-scale general aviation aircraft and rotorcraft. The Air Force provided F-111 crew escape modules and air bags for the tests that were conducted at the facility.

Huey D. Carden and Lisa E. Jones led the F-111 module tests at Langley. The tests (between 60 to 70 drops) spanned from the early 1980's to 1995. The objectives of the tests were to assess

- Impact loads that were generated by the air bag attenuation system
- Structural loads that were transmitted to the seats and occupants
- Module stability under various impact attitudes which could exist due to drift from winds
- Design changes to the air bag attenuation system or seats.

Tests with controlled pitch, yaw, and roll orientations of the module relative to the forward velocity vector were conducted to account for various attitude envelopes of the module during descent. Additional vertical tests were also performed. Impact velocities, structural impact loads, air bag pressures, and loads transmitted to the seats and dummies representing the crew were measured and provided to the Air Force for assessment.

Additionally, impact and postimpact behavior of the module was provided via extensive onboard and ground-based cameras, which also provided module stability information. The last series of drop tests in 1995 assessed the performance of and qualified an entirely new air bag design that was required for a new parachute design.

The results of the Langley tests were analyzed and provided to the Air Force and the industry contractor for continual refinement of the system. As a result of the data provided to the Air Force, load attenuating crew seats were included in the F-111 and air bag and blowout plug design changes were made in the original air bags. Various changes to the module led to the design of a new air bag system, which was tested for qualification on the F-111 in the final series of tests prior to the retirement of the U.S. F-111 fleet.

F-111 crew escape module during tests at the Langley Impact Dynamics Research Facility.

Wing Box Problems

The F-111 airframe utilized a significant amount of high-strength D6ac steel in the wing carry-through structure. This component was heat treated to a tensile strength of 220,000 psi and designed for -3g to 7.33g with design flight life goals of 4,000 hr and 10 years of service. However, a full-scale static test program that was conducted over a 6-year period encountered several failures, including a failure at the wing-pivot fitting. Various modifications, including the first use of an advanced boron-reinforced composite doubler to reduce stress levels, coupled with an extension of the structural tests to 40,000 hr, were believed to have provided for 10,000 hr of safe operations.

In December 1969, an F-111 experienced a catastrophic wing failure during a pull-up from a simulated bombing run at Nellis Air Force Base. This aircraft only had about 100 hr of flight time when the wing failed. The failure originated from a fatigue crack, which had emanated from a sharp-edged forging defect in the wing-pivot fitting. As a result of the accident, the Air Force convened several special committees to investigate the failure and recommend a recovery program. James C. Newman, Jr. and Herbert F. Hardrath represented Langley on the recovery team deliberations, and along with Charles M. Hudson and Wolf Elber, they conducted fatigue crack growth and fracture tests on specimens made from the D6ac steel used in the aircraft. These tests were conducted in the Langley Fatigue and Fracture Laboratory under conditions that simulated aircraft operations. The original material had low fracture toughness due to the heat-treatment process. The committee recommended that every F-111 be subjected to a low-temperature proof test. This proof-test concept had been developed and successfully used in the Apollo program, as well as other missile and space efforts. To screen out the smallest possible flaw size, the F-111 full-scale proof tests were conducted at temperatures of about -40° F, where the fracture toughness of the D6ac steel was lower than the fracture toughness at room temperature. The heat-treatment process was also corrected to provide improved toughness for the D6ac material in newer aircraft. A decade later, the same material with improved toughness was also successfully used in the Space Shuttle solid rocket boosters. As a result of the revised proof-test approach and the improved toughness material, there were no F-111 aircraft lost due to structural failure in almost 30 years of operations before the aircraft was retired from service in 1996.

The F-111 failure was most responsible for the U.S. Air Force developing the damage-tolerant design concept, where flaws, such as a 0.05-in. crack, are assumed to exist in critical aircraft components. The structural components must then be tolerant of these defects during flight conditions. This concept relies on fatigue crack growth and fracture criteria to establish an inspection interval to insure the safety and reliability of the aircraft.

The Transonic Aircraft Technology (TACT) Program

Richard Whitcomb's pioneering research and development efforts on supercritical airfoils for enhanced transonic performance, which began at Langley in 1964 and continued until the 1980's, included extensive wind-tunnel and flight evaluations for potential military applications. After flight tests of a modified F-8 Crusader validated the benefits of supercritical wing technology that had been predicted by theory and wind-tunnel experiments for potential civil applications, NASA and the Air Force became interested in assessing the benefits of supercritical wing applications to high-performance aircraft during transonic maneuvers. Significant increases in the drag-divergence Mach number, the maximum lift coefficient for buffet onset, and the Mach number for buffet onset at a given lift coefficient were demonstrated for the supercritical airfoil when compared with a NACA 6-series airfoil of comparable thickness. Theodore Ayers, in cooperation with General Dynamics, conducted exploratory tests in the Langley 8-Foot Transonic

Pressure Tunnel in 1966 to test the effects of a slotted supercritical airfoil on a 1/15-scale model of the F-111 with the existing flap system. Although the overall results were not satisfactory because of high subcritical drag levels, the results encouraged additional studies of an integral or unslotted supercritical airfoil. Tests of a 1/24-scale F-111 model showed significant benefits to drag-divergence Mach number, maneuver drag, and buffet onset characteristics. These tests also spurred additional interest by General Dynamics in potential improvements of the F-111.

Following these exploratory tests at Langley, interest in supercritical applications continued to increase and a joint NASA and Air Force study that included ground and flight activities was proposed. Specific objectives of the study included the effects of supercritical wings on transonic drag, buffet onset and magnitude, and handling qualities. In a study known as the Transonic Aircraft Technology (TACT) Program, several candidate military aircraft were examined for potential modifications to provide flight validation of supercritical wing military applications.

The program was ultimately based on the application of supercritical technology to the F-111 configuration, which had outer wing panels that could be relatively easily replaced with modified supercritical sections. In addition, the potential retrofit of advanced wings to enhance performance of the F-111 fleet was an interest in some areas of the Air Force. The TACT Program, which started in 1969, included NASA Langley, Dryden, and Ames Research Centers, and the Air Force Flight Dynamics Laboratory. Theodore Ayers served as the Langley focal point for TACT activities, which included extensive experimental development work of the modified wing in the 8-Foot Transonic Pressure Tunnel by Ayers and James B. Hallissy (with considerable oversight and participation from Dr. Whitcomb), tests of a propulsion model in 1973 in the 16-Foot Transonic Tunnel by Charles Mercer, and flutter tests led by Charles Ruhlin and Maynard Sandford in 1971 in the 16-Foot Transonic Dynamics Tunnel.

James B. Hallissy inspects the F-111 TACT model in the Langley 8-Foot Transonic Pressure Tunnel.

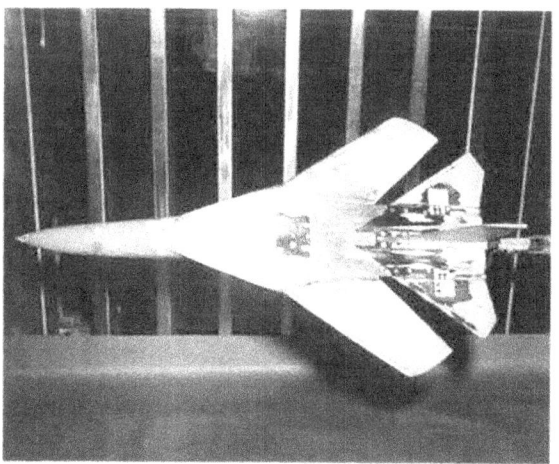

F-111 TACT model in the Langley Unitary Plan Wind Tunnel for supersonic tests.

*F-111 TACT model mounted for flutter tests in the Langley
16-Foot Transonic Dynamics Tunnel in 1971.*

The NASA and Air Force F-111 TACT aircraft during flight tests at Dryden.

The F-111 TACT aircraft began flight tests at Dryden in 1972. Flight-test results showed
that the supercritical wing generated up to 50 percent more lift during maneuvers than
the conventional F-111 wing and significantly delayed the onset of wing buffet to higher
angles of attack. Special flight tests were also conducted to demonstrate that the carriage
of external stores on the wing pylons did not significantly degrade the benefits of the
supercritical wing. Although the Air Force decided not to retrofit the F-111 fleet, the
supercritical wing technology had dramatically demonstrated its benefits for incorporation into future military aircraft.

The Mission Adaptive Wing (MAW) Program

In 1976, Theodore Ayers transferred from Langley to Dryden, where he accepted a position as Director for Aerodynamics. Ayers continued his interest in advanced wing configurations. Working with his technical peers within NASA, DOD, and industry, he advocated for another important flight program with the modified F-111 at Dryden. After the TACT Program ended in the 1980's, the Air Force and NASA engaged in a new technology development program known as Advanced Fighter Technology Integration (AFTI). One element of this joint program was the modification and flight tests of the F-111 TACT aircraft with a Mission Adaptive Wing (MAW). The MAW used flexible wing skin and internal hydraulic control mechanisms to recontour the wing shape as a smooth variable-camber wing for varying flight conditions. The objective was to provide the technical confidence for significant performance improvements with a wing system that varied the wing contour in flight as a function of pilot inputs, flight conditions, and structural loads. The wing box of the existing TACT aircraft was equipped with flexible wing leading and trailing edges. A high-lift section was used for low-speed landing conditions, and the wing was recontoured to a supercritical shape for transonic flight and adjusted to a symmetrical section for supersonic flight. Boeing, Grumman, and General Dynamics bid on the request for proposals for the aircraft modification. Tests of the competing configurations in the 8-Foot Transonic Pressure Tunnel were conducted by James Hallissy. Boeing was awarded the contract in 1979. The MAW Program was managed by the Flight Dynamics Laboratory of the Air Force Wright Aeronautical Laboratories with Dryden as the responsible flight-test organization. Langley supported this activity with exploratory tests of smooth variable-camber concepts in the 8-Foot Transonic Pressure Tunnel, which were conducted by James C. Ferris. Langley also actively participated in the flight-test program, including conducting additional tests with a 1/24-scale model. Langley personnel were also assigned temporarily at Dryden during the flight program. The F-111 MAW flight research was conducted from 1985 to 1988 and included an assessment of automatic camber modes.

The F-111 Mission Adaptive Wing (MAW) aircraft during flight tests at Dryden.

Grumman A-6 Intruder

SPECIFICATIONS

Manufacturer
    Grumman
Date in service
    A-6A 1963
    A-6E 1972
Type
    Attack
Crew
    Two
Engine
    Pratt & Whitney J52-P8B

USERS

U.S. Navy and U.S. Marine
Corps (retired in 1997)

DIMENSIONS

Wingspan . . . . 53.0 ft
Length . . . . 54.7 ft
Height . . . . 15.5 ft
Wing area . . . . .529 sq ft

WEIGHT

Empty . . . . 28,000 lb
Gross . . . . 58,600 lb

PERFORMANCE

Max speed . . . . .563 knots

HIGHLIGHTS OF RESEARCH BY LANGLEY FOR THE A-6

1. Wind-tunnel studies of aerodynamic and thermal characteristics in the Langley 16-Foot Transonic Tunnel provided data for selection of the configuration for the fuselage-mounted speed brakes.

2. Tests of A-6 configurations in the Langley 20-Foot Vertical Spin Tunnel determined that spin recovery characteristics were significantly enhanced by enlarging the rudder and increasing the deflections of the horizontal tail and rudder.

3. Flutter clearance tests for a composite wing for the A-6E were successfully conducted in the Langley 16-Foot Transonic Dynamics Tunnel.

The Grumman (now Northrop Grumman) A-6 Intruder was an all-weather, two seat, subsonic, carrier-based attack aircraft. Designed in the late 1950's, the Intruder played a critical role in the Vietnam War with over 35,000 combat sorties by 1973. The last version of the aircraft, the A-6E, was widely regarded as the best all-weather precision bomber in the world. As an example of its effectiveness, during the strike on Libyan terrorist-related targets in 1986, A-6E Intruders penetrated sophisticated Libyan air defense systems, which had been alerted by the high level of diplomatic tension and rumors of impending attacks. Evading over 100 guided missiles, the strike force flew at low altitude in complete darkness and accurately delivered laser-guided and other ordnance on target. Composite wing replacements and upgraded systems and weapons improvement programs maintained the A-6's combat capability until its retirement from the fleet in early 1997.

Langley's contributions to the A-6 Program began in 1959. Studies of the aerodynamic and thermal characteristics of candidate designs for the fuselage-mounted speed brakes, which were located adjacent to the hot efflux of the turbojet engines, were conducted in the Langley 16-Foot Transonic Tunnel. Tests were also conducted in the 20-Foot Vertical Spin Tunnel to determine the spin and recovery characteristics of the aircraft and to determine the size of parachute required for the spin test aircraft. Initial Langley tests indicated that spin recoveries were significantly improved with an enlarged rudder and more deflection angles for the horizontal tail and rudder. The rudder was enlarged, and the control system was modified to provide the extended deflections. Follow-on spin tests were also conducted over the years to determine the effects of new external stores and new weapons on spin recovery characteristics.

Arguably, the most important contribution of Langley to the A-6 Program was a series of two test entries in the Langley 16-Foot Transonic Dynamics Tunnel (TDT) to ensure flutter clearance for an advanced composite wing that was incorporated into the A-6E fleet to extend the fatigue life and capabilities of the aircraft. Initial tests in the TDT ended in flutter failure of the model wing, but a revised wing design passed flutter demonstration requirements and permitted the fleet to utilize the advanced wing.

LANGLEY CONTRIBUTIONS TO THE A-6

Speed Brake Studies

In May 1959, a team from Langley and Grumman led by Langley researcher Charles E. Mercer conducted powered-model tests in the Langley 16-Foot Transonic Tunnel to determine the optimum configuration for fuselage-mounted speed brakes–thrust spoilers for the A-6 (then designated the A2F-1). The speed brake panels were located immediately behind the engine exhaust nozzles on the rear fuselage, which caused concern over the aerodynamic loads and the thermal environment of the speed brakes during high-power conditions. To further complicate the engineering issues, the original aircraft also had a tilting tailpipe to increase lift at low speeds and thereby reduce approach speeds. The test examined several candidate configurations, including a perforated panel design, which was subsequently incorporated into the early A-6A and EA-6A. Later variants of the A-6 used symmetrically deflected split ailerons at the wingtips for speed brakes, and the original speed brake panels were inactive or deleted.

The A-6 prototype made its first test flight in April 1960. The Navy was satisfied with the 90-knot approach speed, so the deflectable nozzle design feature was eliminated in production aircraft. The Langley staff, however, continued research on thrust vectoring, which later contributed to more advanced fighter designs.

Powered-model tests in the Langley 16-Foot Transonic Tunnel evaluated several speed brake configurations.

Spin Tunnel Tests

Initial evaluations of the spin and spin recovery characteristics of the A-6 by Henry A. Lee in the Langley 20-Foot Vertical Spin Tunnel indicated that spin recoveries could be significantly enhanced by an enlarged rudder and by additional deflections of the rudder and horizontal tail. The original A-6 design had incorporated extended throws on the rudder and horizontal stabilizer for the power approach configuration, so an assist spin recovery switch was installed so the pilot could extended deflections for the cruise configuration. The Langley tests determined that the greater horizontal-tail deflection angle (trailing-edge up) provided a better flow field on the vertical tail during spins, thereby increasing the effectiveness of the rudder to terminate the spin. The rudder of the production aircraft was enlarged as a result of these tests.

The Composite Wing

By the mid-1980's, the A-6 was beginning to show its age. The accumulated stress of high-g catapulted takeoffs and arrested landings on carriers and the long exposure to salt water were beginning to take a toll on the life of airframe components. Studies were underway for a new wing design, and inspections of the A-6 structure revealed major corrosion problems. In January 1987, a fatal accident resulted in an investigation of the structural health and projected lifetime of the A-6 fleet (ref. 5).

At the time of the accident, U.S. Marine Corps First Lieutenant Bob Pandis and his bombardier/navigator, Lieutenant Colonel John Cavin, were practicing dive-bombing missions at the El Centro Naval Base in California. During a 40-deg diving run at about 500 knots, the left wing of their A-6E broke off the aircraft at an altitude of about 8,000 ft and the aircraft began to spin wildly out of control. When the wing separated from the aircraft, fuel from the severed fuel cells in the aircraft immediately ignited in a giant fireball, adding to the severity of the rolling motion. The violence of the roll caused the empennage to separate from the aircraft, which started cartwheeling end-over-end. Pandis ejected from the out-of-control aircraft and suffered major injuries, while Cavin died in the crash.

Pandis later recalled that the two other A-6's in his flight had noticed fuel venting from his left wing during the run for the dive. About halfway into the dive, the crews of the two aircraft saw a dramatic increase in fuel venting and were shocked to see the left wing tear off the aircraft.

Prior to this terrible accident, 72 A-6 aircraft had been temporarily grounded and another 109 were operating under flight restrictions. The Boeing Company had begun to design a new wing under a Navy contract that was awarded in 1985. With a

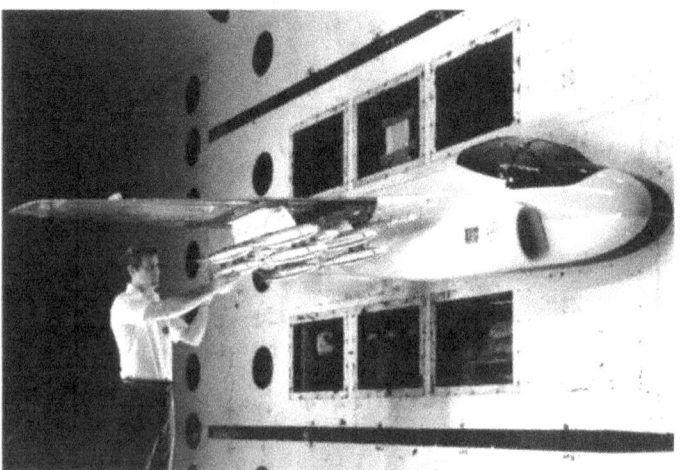

Flutter tests of the new A-6 composite wing in the Langley 16-Foot Transonic Dynamics Tunnel included numerous external store loadings.

carbon fiber–epoxy resin torsion box, light alloy control surfaces, and some titanium components, the new wing was much lighter and designed for four times the fatigue life of the existing wing. In addition to becoming a retrofit for the A-6E fleet, the wing was also intended for a new, advanced version of the A-6 to be known as the A-6F. (The A-6F was later canceled in the prototype stage when the Navy decided to replace the A-6 fleet with the stealth A-12 aircraft).

At the request of the Navy, tests were conducted in the Langley 16-Foot Transonic Dynamics Tunnel (TDT) in February 1986, to ensure that the new wing would not exhibit flutter within the flight envelope. The preflight flutter clearance tests required two separate entries in the TDT under the leadership of Langley researchers Stanley R. Cole and Jose A. Rivera, Jr. The first composite-wing model was lost in a catastrophic flutter event while testing a "pencil" store configuration that represented the fuel tanks. Pencil stores are slender metal rods that simulate the moments of inertia of the real fuel tanks while minimizing aerodynamic effects. The test of the pencil store configuration is part of the flutter clearance test; however, the result was much more severe flutter than expected and the model was destroyed. Based in part on these results, Boeing modified the structural design of the A-6 composite wing. A second model incorporating the new wing design was fabricated and tested in the TDT in June 1987. This second model demonstrated no flutter incidents within the scaled flight envelope plus a safety margin.

A unique aspect of both models was that they were semispan models (only one half of the aircraft was modeled); however, Boeing designed a unique root constraint to simulate the carry-through structure of the wing at the fuselage. The semispan model was larger than a full-span model, thereby permitting more accurate representation of the structural characteristics at model scale. Also of interest with regard to model design, the first model had a very clean wing surface. However, the decision was made to include the bumps, fences, and other wing features that were present in the actual flight hardware on the second model.

Subsequent to the Langley tests, the composite wing was retrofitted to about 200 A-6E aircraft, which significantly increased the aircraft's capability, safety, and operational life.

By 1988, the team of McDonnell Douglas and General Dynamics began to build the A-6 replacement called the A-12 Avenger. Unfortunately, Defense Secretary Richard Cheney cancelled the $57 billion project after cost overruns exceeded $2.7 billion dollars in the development phase. Nonetheless, the decision to retire the A-6 was unchanged, and the F/A-18 Hornet became the replacement aircraft for the Intruder. The last flight of the A-6 Intruder occurred in early 1997.

Grumman EA-6B Prowler

SPECIFICATIONS

Manufacturer
  Grumman
Date in service
  1971
Type
  Electronic warfare
Crew
  Four
Engine
  Pratt & Whitney J52-P-408A

USERS

U.S. Navy and U.S. Marine
Corps

DIMENSIONS

Wingspan . . . . 53.0 ft
Length . . . . 59.8 ft
Height . . . . 16.3 ft
Wing area . . . . 529 sq ft

WEIGHT

Empty . . . . 34,300 lb
Gross . . . . 61,500 lb

PERFORMANCE

Max speed . . . . . Mach number
of 0.72

HIGHLIGHTS OF RESEARCH BY LANGLEY FOR THE EA-6B

1. Tests in the Langley 20-Foot Vertical Spin Tunnel resulted in increased control surface deflections for spin recovery.

2. Langley cooperated with the Navy and Grumman to determine configuration modifications that significantly improved maneuver aerodynamics, high-angle-of-attack characteristics, and low-speed high-lift capability.

3. Aft-fuselage ram-air cooling scoops were designed with data from tests in the Langley 7- by 10-Foot High-Speed Tunnel.

The Grumman (now Northrop Grumman) EA-6B Prowler is a unique national asset that can be deployed from land bases and aircraft carriers to monitor the electromagnetic spectrum and actively deny an adversary the use of radar and communications. This aircraft is included in every aircraft carrier deployment and has a primary mission to protect fleet surface units and other aircraft by jamming hostile radar and communications. The EA-6B played a key role in suppressing enemy air defenses during Operation Desert Storm for U.S. and allied forces. The Department of Defense (DOD) has now assigned all radar jamming missions to the Prowler.

The EA-6B airframe is a derivative of the A-6E Intruder. Although the fuselage was stretched to permit space for four crew members, the wing area is the same as that used on the A-6E. The basic mission, external stores, and electronic suites of the Prowler are considerably different from those of the A-6E. Also, the EA-6B is considerably heavier than the A-6, which results in a significant reduction in maneuvering capability. As a result of an alarming increase in accident rates for the EA-6B fleet in the early 1980's, the Navy requested support from Langley to define modifications that might improve maneuver aerodynamics, high-angle-of-attack stability and control, and low-speed high-lift systems. During follow-on Navy flight tests, these modifications to the wing airfoil, vertical tail, wing leading and trailing edges, and roll control devices significantly enhanced the capabilities of an EA-6B demonstrator aircraft. Langley also supported a Navy and Grumman request to conduct wind-tunnel tests to define fuselage ram-air cooling scoop concepts for additional aerodynamic cooling of electronic systems. The significant improvements predicted by the results of Langley tests were verified during Navy flight evaluations of a modified A-6F aircraft.

Langley facilities that supported the EA-6B program included the 20-Foot Vertical Spin Tunnel, the National Transonic Facility (NTF), the 7- by 10-Foot High-Speed Tunnel, the 16-Foot Transonic Tunnel, the 30- by 60-Foot (Full-Scale) Tunnel, the Low-Turbulence Pressure Tunnel (LTPT), the 6- by 28-Inch Transonic Tunnel, and the 12-Foot Low-Speed Tunnel.

LANGLEY CONTRIBUTIONS TO THE EA-6B

Spin Tunnel Tests

The EA-6B aircraft is a four-seat, electronic warfare derivative of the A-6E Intruder. The major external differences between the EA-6B and the two-seat A-6E are a 54-in. fuselage extension for two additional crew stations, a large pod on the vertical tail to house electronic countermeasures equipment, and a canted refueling probe. The EA-6B is more than 10,000 lb heavier than the A-6E, and carries large pods for electronics on wing pylons. In view of these significant configuration changes, the Navy requested that Langley conduct tests to determine spin and recovery characteristics of the EA-6B. Henry A. Lee and James S. Bowman, Jr. conducted the investigation in 1971.

As discussed in *Langley Contributions to the A-6*, a cockpit switch had been implemented in the A-6 to provide the pilot with the option of increasing horizontal-tail and rudder deflections for spin recovery. The results of the tests of the EA-6B indicated that the existing 23-deg rudder deflection was not sufficient for satisfactory spin recovery. The rudder deflection on production aircraft was increased to 35 deg with the A-6 assist spin recovery switch, which was also implemented on the EA-6B.

Maneuver Improvements

The EA-6B has a significantly higher design gross weight than the A-6E; however, the EA-6B employs the same wing to carry the increased load. This increased wing load contributed to an alarming number of EA-6B accidents in the early 1980's. During that period the EA-6B aircraft experienced accident rates in fleet operations that were nearly three times higher than all other Navy and Marine aircraft combined. The majority of these mishaps were attributed to out-of-control flight and resulted in the loss of the aircraft after the pilots were unable to recover the aircraft and were forced to eject. These losses prompted many fleet squadrons to restrict the EA-6B from intentional maneuvers at high-angle-of-attack conditions. While the restrictions substantially reduced the accident rates, they also imposed constraints on evasive maneuvers while operating in high threat environments.

In late 1984, the Navy approached Langley to undertake a research program to improve the EA-6B, with emphasis on increasing maximum usable lift, maintaining lateral-directional stability near stall, and maintaining lateral control near stall. Langley agreed to lead this effort, under the cognizance of the Navy. Grumman joined the effort and provided additional technical support to Langley, as well as participating in additional wind-tunnel tests at NASA Ames Research Center for the high-lift configuration. This program evaluated the EA-6B in ten NASA and Grumman facilities.

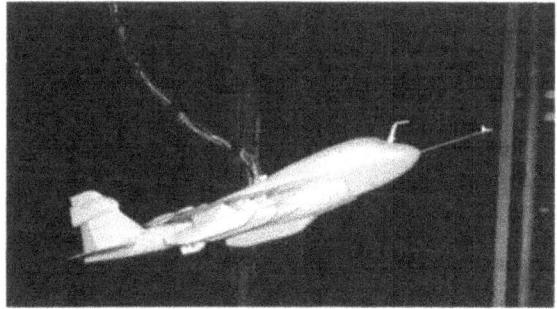

The modified EA-6B model in flight in the Full-Scale Tunnel.

Langley researcher David E. Hahne with the EA-6B free-flight model.

Langley research activities on the high-angle-of-attack stability and control characteristics of the EA-6B were led by Joseph L. Johnson, Jr. and his staff of the Low Speed Aerodynamics Division. The efforts to improve maximum lift and maneuver performance were led by Percy J. (Bud) Bobbitt and his staff of the Transonic Aerodynamics Division.

Johnson's team conducted extensive tests in the Langley 12-Foot Low-Speed Tunnel and the Langley 30- by 60-Foot (Full-Scale) Tunnel with flow visualization and force and moment measurements. The Langley leaders in this effort were Frank L. Jordan, Jr. and David E. Hahne. The results of these tests showed that the vertical tail was adversely affected by flow emanating from the fuselage and wing root areas for high-angle-of-attack conditions, which resulted in a severe loss of directional stability. The problem was resolved by extending the vertical tail above the existing fin pod, adding leading-edge droop to the inboard wing, and adding a strake to the wing-fuselage intersection. Roll control at high-angle-of-attack conditions was augmented by using the existing wingtip speed brakes as additional ailerons. These modifications were tested with a 0.12-scale free-flight model in the Full-Scale Tunnel. The results clearly demonstrated the potential of these modifications to improve the flight characteristics of the EA-6B at high angles of attack.

Bobbitt's team worked to improve maximum lift. This task was a challenge as changes in wing contour were limited to the leading-edge slat and trailing-edge flap to keep retrofit costs low. Several advanced airfoil designs were tested by William G. Sewall, Robert J. McGhee, and James C. Ferris in the Langley 6- by 28-Inch Transonic Tunnel and in the Langley Low-Turbulence Pressure Tunnel (LTPT). The results of the airfoil studies were used to design advanced slats and flaps for the EA-6B that produced a substantial increase in lift, as well as decreased drag at cruise Mach numbers. Edward G. Waggoner and Dennis O. Allison led computational and experimental studies to design the wing configurations to increase the low-speed maximum lift capability of the EA-6B with minimal degradation in high-speed performance. Their work ranged from the application of low-speed and transonic computational methods to experimental verification tests in the Langley 7- by 10-Foot High-Speed Tunnel and the National Transonic Facility (NTF) at Langley. The integrated efforts of these two teams identified new leading- and trailing-edge configurations that dramatically improved the high-lift performance of the EA-6B. The modifications defined by the Johnson and Bobbitt teams were then tested on a model at Ames to assess the impact on the high-lift configuration. Spin tunnel tests were also conducted to evaluate the impact of the changes on spin and spin recovery characteristics. No degradation was noted during the tests, and the existing spin characteristics of the EA-6B were projected to remain unchanged. The final tests included wing loads tests led by Langley researcher Charles Mercer in the Langley 16-Foot Transonic Tunnel. Tests of a full-size, EA-6B semispan wing in the Full-Scale Tunnel led by Jordan and Hahne validated the effectiveness of using the speed brakes as ailerons and provided additional loads data. Because the A-6E had significant problems with fatigue life, it was important to assess the impact of the modifications on wing loads and wing root bending moments to ensure that the EA-6B wing fatigue life would not be degraded.

The impact of the Langley-led test program was substantial. Test data indicated the potential for a 25-percent increase in maximum usable lift in the cruise configuration. Lateral and directional stability could be maintained to angles of attack well beyond stall. Lateral control could be maintained beyond stall by using the speed brakes as ailerons. A performance improvement due to decreased drag could be realized at medium

EA-6B model being prepared for tests in the National Transonic Facility at Langley.

*EA-6B Vehicle Enhancement Program (VEP)
test aircraft during flight evaluation.*

and high altitudes where a majority of the EA-6B missions are flown. Approach speeds could be substantially reduced at existing landing gross weights, and growth capability was provided for higher gross weights. Finally, an extensive, unprecedented aerodynamic database for the EA-6B had been produced and was ready for incorporation into simulators for EA-6B pilot training.

The results of this outstanding NASA, Navy, and Grumman joint program were summarized in several technical papers in 1987. The effort received such positive national attention that in the 1987 Applied Aerodynamics AIAA meeting, an entire session was dedicated to the EA-6B effort with five papers presented. Aircraft modifications defined by the joint EA-6B program were directed at an advanced version of the aircraft known as the EA-6B ADVCAP (Advanced Capability), which included a myriad of improvements such as new engines and electronic upgrades, as well as the airframe modifications. Flight tests of a modified EA-6B designated the Vehicle Enhancement Program (VEP) test aircraft were conducted to evaluate the effects of the airframe modifications derived from the Langley-led joint studies. Flights began on June 15, 1992, and the final flight occurred on April 4, 1994. The flight evaluation of the test aircraft validated all projections of the research program, and the performance, stability, and control characteristics of the modified aircraft were judged to be far superior to those of the basic EA-6B. The aircraft performed flawlessly. Unfortunately, fiscal constraints and other programmatic issues restrained the Navy from funding the upgrades. EA-6B simulators, however, have been upgraded with the data from Langley.

Cooling Study

The continuing evolution of the EA-6B and the demands of the all-weather warfare mission involved a continuous upgrade of advanced electronic systems that demand adequate cooling for satisfactory operations. In particular, the projected Advanced Capability (ADVCAP) version of the aircraft included a significant electronics upgrade that required additional cooling. Langley participated in a joint NASA, Navy, and Grumman wind-tunnel study of new aft-fuselage ram-air scoops to satisfy the cooling requirements. Langley researcher William Sewall led these tests, which were conducted in the Langley 7- by 10-Foot High-Speed Tunnel. The scoop configuration derived from the tests was evaluated in flight tests of a prototype of another advanced A-6 derivative known as the A-6F. The scoop configuration was to be included in the overall ADVCAP package, which was not funded for continued development.

EA-6B fuselage and empennage model in the Langley 7- by 10-Foot High-Speed Tunnel for air scoop studies.

A-6F prototype with aft-fuselage scoops.

Grumman F-14 Tomcat

SPECIFICATIONS

Manufacturer
  Grumman
Date in service
  November 1974
Type
  Carrier-based multirole
Crew
  Two
Engine
  F-14B/D . . . General Electric
  F-110-GE-400 augmented
  turbofan with afterburner

USER

U.S. Navy

DIMENSIONS

Wingspan
  unswept . . . . 64.2 ft
  swept . . . . 38.2 ft
Length . . . . 61.9 ft
Height . . . . 16.0 ft
Wing area . . . . 565 sq ft

WEIGHT

Empty . . . . 41,700 lb
Max takeoff . . . . 74,349 lb

PERFORMANCE

Max speed . . . . . Mach number
                        of 2.3
Unrefueled
range . . . . 3,450 n mi

HIGHLIGHTS OF RESEARCH BY LANGLEY FOR THE F-14

1. The F-14 employs the variable-sweep wing developed by Langley to meet wide ranging mission requirements.

2. Langley conducted wind-tunnel tests to collect data for the competitive source selection team.

3. Langley personnel participated in the evaluation process.

4. At the request of the Department of Defense, Langley led a multidiscipline assessment of the F-14 in early development and reported the results to the Navy.

5. Langley conducted propulsion integration studies to improve the subsonic cruise efficiency that resulted in a redesign of the aft end.

6. Langley studies of the high-angle-of-attack and spin characteristics identified a potential flat spin and an innovative flight control system modification that dramatically reduced the spin susceptibility of the aircraft.

7. Tests in the Langley 16-Foot Transonic Dynamics Tunnel identified potential flutter of the F-14 over-wing covers, which Grumman solved with external strake stiffeners.

8. Langley provides technology, expertise, and facility support on a continual basis, such as providing numerical models for pilot training simulators.

The multirole F-14 fighter employs many Langley technical concepts that permit it to accomplish diverse requirements such as supersonic dash and landing on an aircraft carrier in adverse conditions. Grumman relied on existing NASA databases and consultation during the design of the F-14. Langley staff tested and analyzed the competing designs in the Navy Advanced Fighter (VFX) Program competition that resulted in the F-14. Technical contributions to the F-14 from Langley include the areas of aerodynamic performance, high-angle-of-attack and spin characteristics, and flutter suppression. Langley facilities used in F-14 studies included the 30- by 60-Foot (Full-Scale) Tunnel, the 20-Foot Vertical Spin Tunnel, the 4- by 4-Foot Supersonic Pressure Tunnel, the 8-Foot Transonic Pressure Tunnel, the Unitary Plan Wind Tunnel, the 16-Foot Transonic Tunnel, the 16-Foot Transonic Dynamics Tunnel, the 7- by 10-Foot High-Speed Tunnel, the Differential Maneuvering Simulator, and radio-controlled models dropped from a helicopter. In addition, Langley provided leadership for high-angle-of-attack F-14 flight experiments in a joint NASA, Navy, and Grumman program at NASA Dryden Flight Research Center.

Following the selection of the Grumman design at the completion of the VFX Program, the Department of Defense requested that NASA provide an independent assessment of the F-14. A NASA team of 45 professionals that was led by Langley briefed their findings to the Navy and helped shape the development of the aircraft. Over 11,000 hr of wind-tunnel tests were conducted in Langley facilities during the subsequent F-14 development program.

LANGLEY CONTRIBUTIONS TO THE F-14

Early NASA Fighter Studies

During the development of the variable-sweep wing at the Langley Research Center, researchers recognized the advantages of applying the concept to multimission aircraft. One ideal application was for naval fleet defense fighters, which must be able to quickly intercept threats and yet slowly approach aircraft carriers to land. Variable-sweep wings in the fully swept (high-speed) configuration permit efficient supersonic dash and the carrier-approach requirements could be met with the wing in the unswept (low-speed) position. Inspired by the potential of this application of variable-sweep technology, Langley conducted several in-house studies of fighter configurations for naval applications. In 1967, Langley published the results of in-house studies of a variable-sweep fighter configuration known as LFAX-4 that incorporated several features that are evident in the F-14 aircraft. Some of these features are

- Short propulsion package to minimize weight
- Engines placed forward for balance
- Horizontal ramp engine inlets for good performance at high angles of attack
- Tailored twin-engine aft-end spacing and interfairing for efficient subsonic cruise conditions

Results of studies of the performance, stability, and control characteristics of the LFAX-4 obtained across the speed range of a representative advanced naval fighter were outstanding, and Langley researchers conducted extensive briefings of their findings at industry and Department of Defense (DOD) sites.

One result of the LFAX-4 study that the Langley team emphasized was the critical location of the pivot for the movable wing panels. To minimize drag during transonic

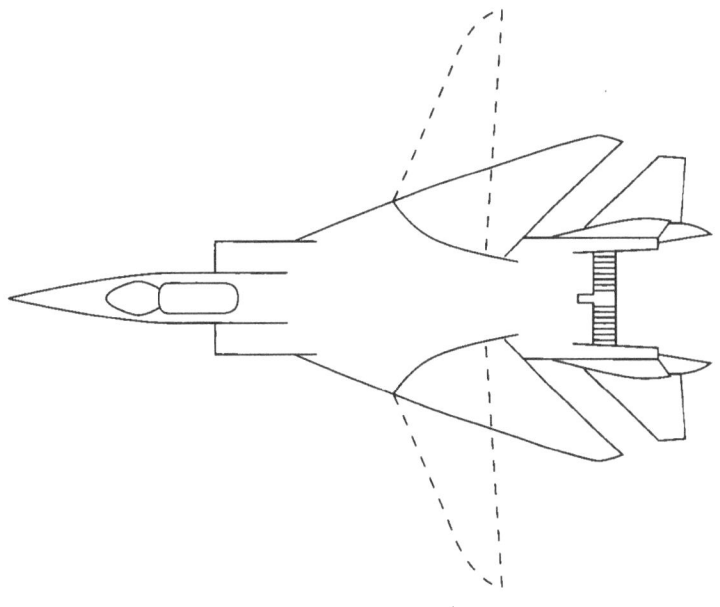

The Langley LFAX-4 fighter configuration.

maneuvers typical of air-to-air combat, the pivots must be located in a relatively outboard position. Langley's experience with the test and analysis of the F-111 revealed that large penalties in trim drag occurred if this key design factor was not adequately appreciated. Although the F-111 incorporated the variable-sweep concept, the full advantages of the concept were not realized because the pivot locations were relatively inboard. As a result, the F-111 suffered excessive trim drag at transonic and supersonic conditions. The designers of the F-14 were made aware of the significance of pivot locations by NASA briefings. Comparison of the NASA results for the LFAX-4 to those of the F-111 helped convince Grumman to locate the F-14 pivots in a more favorable outboard position.

The LFAX-4 studies also pointed out the importance of the placement of the horizontal tail relative to the wing for stability and control. These recommendations were also included in the F-14 design.

F-14 Source Selection

Following the disastrous attempt to achieve interservice aircraft commonality with the F-111, the Navy issued a request for proposals (RFP) for a new VFX fighter in July 1968. Competitors included Grumman, General Dynamics, Ling-Temco-Vought, McDonnell Douglas, and North American Rockwell. At the request of the Navy, the competing configurations were evaluated with wind-tunnel data from the Langley 8-Foot Transonic Pressure Tunnel, the 7- by 10-Foot High-Speed Tunnel, the Unitary Plan Wind Tunnel, and the 16-Foot Transonic Tunnel. Langley personnel also participated in briefings of their results to the industry teams and the Navy.

NASA Independent Assessment

On January 14, 1969, the Navy announced the award of the contract for the VFX fighter, now designated F-14, to Grumman. The Assistant Secretary of the Navy requested that NASA make a timely independent assessment of the technical development of the F-14. A NASA F-14 Study Team of over 40 Langley personnel led by Langley researcher William J. Alford, Jr. was organized. A briefing of the study results was given in August 1969 at the Naval Air Systems Command by a team led by Laurence K. Loftin, Jr., Mark R. Nichols, and William Alford. This briefing (which covered results in cruise and maneuvering performance, aeroelasticity and flutter, propulsion integration, stability, and control) identified several areas where further research would enhance the F-14's capabilities. Following the briefing, Dr. John Foster, Director of Defense Research and Engineering, requested the support of Langley in the development of the F-14.

Propulsion Integration

Initial flight-test evaluations of the performance of the F-14 by the Navy revealed higher drag levels at high subsonic and transonic speeds than had been predicted by Grumman wind-tunnel tests. The Navy requested Langley support in analyzing and providing solutions to the problem. Langley experience with the F-111 and other advanced fighter concepts indicated that an extremely large portion of the high subsonic and transonic cruise drag of modern twin-engine fighters is contributed by the aft end of the configuration. For example, a relatively poor aft-end design could produce almost 50 percent of the cruise drag for some configurations. In the course of the F-111 development program, Langley researchers in the 16-Foot Transonic Tunnel had developed test techniques and analysis methods to minimize this problem, and they went to work on the aft-end aerodynamic characteristics of the F-14 configuration.

Based on their extensive experience, the 16-Foot Transonic Tunnel team conducted tests in 1972 to determine the characteristics of the critical engine-fuselage fairing (pancake) at the rear of the aircraft. Several geometric variations were evaluated to determine more effective pancake shapes, with an appreciation of the trades that are necessary to minimize component interference drag while adhering to the area rule developed by Langley researcher Dr. Richard Whitcomb. In addition, certain regions had to be preserved for the F-14 fuel jettison system and the landing arrestor hook. The researchers cut away areas of the pancake, reshaped the geometry and added a "handle" in the shape of a bulbous pod at the rear of the pancake. The design recommended by the Langley tests proved extremely effective in reducing cruise drag and was incorporated into the F-14 configuration.

In addition to the pancake modification, Langley researchers recommended that generous "speed bump" (area added to shape the aircraft to comply with Whitcomb's area rule) fairings be added to the forward bottom area of the vertical tails. The suggestion was accepted by Grumman and incorporated into the production F-14 fleet.

Pancake configurations evaluated in cruise-drag studies in the Langley 16-Foot Transonic Tunnel. Original shape is at left.

Aft-end pancake handle and vertical tail root speed bumps on F-14 aircraft.

High-Angle-Of-Attack and Spin Research

In early 1970, initial tests conducted in the Langley 20-Foot Vertical Spin Tunnel at the request of the Navy indicated that the F-14 would exhibit two types of spins. The first spin involved relatively steep, nose-down spins from which recovery would be relatively easy for the pilot. However, the results also showed that the F-14 might exhibit a relatively flat unrecoverable spin in which the aircraft would rotate rapidly (2 sec per turn) about a vertical axis through its center of gravity, while descending vertically with the fuselage in a relatively horizontal attitude. Because of the high rate of rotation of the flat spin, the g-forces at the cockpit location would be very high (approximately 6.5 longitudinal g's outward) and would probably incapacitate the pilot if sustained for even a moderate period of time. Under the direction of Langley researcher James S. Bowman, Jr. exhaustive spin tunnel tests were conducted to define a spin recovery procedure, but the aerodynamic control surfaces of the F-14 were ineffective at these flat attitudes. In fact, even a scaled version of a very large 35-ft diameter spin recovery parachute (the largest that could be carried by the F-14 was 21 ft) could not recover the model from the spinning motions. The only concept that provided marginal recovery was the simultaneous application of normal recovery controls, deployment of the emergency parachute, and extension of auxiliary canards on the nose of the model.

Unrecoverable flat spins have been exhibited by many fighter configurations in the Langley Spin Tunnel, and such characteristics are viewed with concern. However, additional types of model tests are required to judge the seriousness of the problem. In particular, drop-model tests are conducted to determine if the aircraft can enter the spin from initial flight conditions. Spin tunnel tests are conducted with the model launched in a flat attitude into a vertically rising airstream—conditions very favorable for the spin to stabilize. However, in actual flight conditions, many aircraft lack the control power required to reach these conditions. For example, although the F-5 aircraft exhibited a flat spin in Langley Spin Tunnel tests, it was virtually impossible for pilots to intentionally enter the flat spin.

F-14 model in spin recovery tests in the Langley Spin Tunnel.

F-14 drop model being calibrated in flight electronics lab prior to flight tests.

Two F-14 drop models were under fabrication when the spin tunnel results became known. The Langley, Grumman, and Navy team had planned to equip the models with extensive instrumentation to measure flight variables for correlation with analytical studies of the spin, but the installation process would have taken several months. Because the Navy required an immediate answer about the susceptibility of the F-14 to enter flat spins, Marion O. McKinney directed the team to install limited instrumentation in one of the models, and to obtain answers as quickly as possible. With the approval of this approach by the Navy, Charles E. Libbey and his team installed a miniature movie camera in the engine inlet of the model and pointed it forward, where it could monitor the relative angle of a simple wooden angle-of-attack vane mounted on a nose boom. This innovative approach gave the Navy answers in a few weeks, rather than months. The team conducted as many as 4 drop tests a day in a very successful test series. More refined instrumented model tests were completed later for a total of 55 drop tests in the program.

Libbey and his team concentrated their initial studies on the susceptibility of the F-14 model to enter the spin by manipulating the controls after the model had been dropped from a helicopter. Results of the tests showed that the model could be pitched up in an aggressive manner with no tendency to enter either the steep or flat spins. However, if the roll control (differential deflection of the horizontal tails) was used in normal fashion to pick up a down-going wing at high angles of attack, the model would depart controlled flight in a direction opposite of the intended input because of adverse yaw caused by large yawing moments produced by the horizontal tails. If the pilot held in the roll control, the model would enter a flat spin. Recovery from flat spins requires the use of an emergency parachute, special nose canards, and full differential tail deflections.

Although the drop-model results predicted that entry into the flat spin was possible, it was recognized in this test technique that the ground-based pilot controls the model from a remote position, thereby losing the natural physical cues available to the pilot of the actual aircraft. In addition, the scaled model moves much faster than the full-scale aircraft, so the time available to the pilot of the model for recovering from an impending spin is very limited. The analysis of the F-14 therefore turned to the use of the Langley Differential Maneuvering Simulator (DMS) where the actual human environment could be more accurately simulated.

In 1972, a team headed by Langley researcher Luat T. Nguyen programmed the DMS with aerodynamic data from Langley wind-tunnel test results and F-14 aircraft mass characteristics. Arrangements were made for Langley research pilots and the Grumman chief test pilot Charles Sewell to participate in the evaluation. Simulated maneuvers flown in the DMS confirmed the drop-model results—that is, the aircraft could be aggressively pitched to extreme attitudes without loss of control, but if roll control was maintained at high angles of attack, the aircraft would depart from controlled flight. In addition, with simulation of full-scale conditions, it appeared that the pilot had time to make corrective controls before hazardous spin-entry conditions occurred.

Nguyen and his team had participated in the development of the flight control system for the F-15 for high-angle-of-attack conditions. (See *Langley Contributions to the F-15*.) They recognized that the concept used by the F-15 to reduce the adverse effects of the horizontal tails as roll control devices at high angles of attack would be an ideal solution to the F-14 problem. With this knowledge, an automatic rudder interconnect (ARI) for the F-14 was implemented and evaluated in the DMS. The ARI system automatically phased out movement of the tails for roll control and phased in deflections of the rudders at high angles of attack. The concept was refined and matured in the simulator studies. The pilots who flew the F-14 with the ARI system were enthusiastic, and the system allowed the pilot to maneuver the aircraft without regard to angle of attack or switching from differential tails to rudders. The Grumman team regarded the ARI concept developed by Langley as a highly desirable addition to the F-14 aircraft. The production contract for the early F-14 aircraft called for the implementation of an ARI system. At this point, Langley closed out its F-14 research, while Grumman pursued the development of the final ARI system.

Unfortunately, the early F-14 aircraft also included another late developing preproduction concept—deployable wing leading-edge maneuver slats for improved maneuverability. Early Grumman flight tests revealed that the F-14 modified with both the ARI system and the maneuver slats displayed unsatisfactory air combat maneuvering characteristics because the ARI rudder inputs aggravated lightly damped rolling oscillations (wing rock) induced by the slats during maneuvers. Because of this incompatibility, the Navy deactivated the ARI systems on all fleet F-14 aircraft.

The F-14 proved to be a relatively forgiving aircraft to fly, and pilots adapted to manually switching from using differential tails for roll control at low angles of attack to using rudders at high angles of attack. However, the F-14 fleet began to experience spin losses at the rate of about one aircraft per year. In 1978, a joint NASA, Navy, and Grumman program was initiated to develop a new ARI system to increase the spin resistance of the F-14. Nguyen and his team once again examined the F-14 and used the DMS to develop a new ARI that provided adequate damping of the wing rock, while retaining the spin resistance of the original ARI system developed by Langley. A flight-test F-14 was modified with a spin parachute, battery driven hydraulic pumps for emergency power, and the special foldout canards on the fuselage forebody that were recommended by the earlier spin tunnel tests. Fitted with the new ARI, flight tests were conducted over a 2-year period at NASA Dryden Flight Research Center with Langley personnel on site for the flight tests. Over 100 flights by 9 pilots were made up to low supersonic speeds.

The results of the flight-test program were extremely impressive. Wing rock was suppressed, inadvertent spins were eliminated, and the handling qualities throughout the air combat envelope were improved. Several years passed before funding constraints permitted the Navy to develop the ARI within plans to equip the F-14 fleet with a new advanced digital flight control system (DFCS). Following further refinements during Navy flight evaluations at Patuxent River Naval Air Station in Maryland, the Navy implemented the DFCS with the ARI. The first F-14 deployments with the ARI occurred during the Kosovo operations, and glowing reports from the F-14 squadrons indicated that the new system was a success.

The F-14 responds to an abrupt stick pull by Dryden test pilot Fitz Fulton during evaluations of the Langley-designed Automatic Rudder Interconnect (ARI) control system at Dryden.

Flutter Tests

Flutter clearance tests of the F-14 in the Langley 16-Foot Transonic Dynamics Tunnel required five entries from 1970 to 1973 under the leadership of Moses G. Farmer. The utilization of variable-sweep wings by the F-14 introduced a unique flutter problem that had been unanticipated prior to the tests. The challenge of providing an upper-fuselage covering for the variable-sweep wing panels had been addressed by Grumman with relatively flexible inner wing covers. Early in the flutter tests, large deflections and buffeting of the over wing panels were observed and viewed as a potentially serious flutter problem.

With knowledge of the Langley results, Grumman engineers designed a set of external stiffening strakes for the wing covering that eliminated the flutter problem. An additional favorable impact of the strakes was local straightening of the airflow over the upper fuselage, which resulted in performance benefits. With the strake modification, the F-14 passed flutter clearance tests in the Langley tunnel. Initially, it was proposed that this modification would only be applied to the preproduction flight-test F-14 aircraft while a redesign of the wing cover could be accomplished. However the modification proved to be extremely robust and similar strakes were incorporated in all production F-14 aircraft.

During the flutter tests, the Langley staff observed considerable buffeting of the vertical tails, particularly at moderate angles of attack. The staff of the 16-Foot Transonic Dynamics Tunnel modified the cable-mount system to permit tests at high-angle-of-attack conditions where the buffeting became more intense. Langley expressed concern that damage to the structural integrity of tails or tail-mounted avionics and antennae might be encountered, but Grumman did not accept this concern as an issue. Subsequently, the F-14 fleet experienced structural damage and the replacement of tail-mounted radar warning units. As a result, the vertical tails of F-14 aircraft were stiffened.

The experience of the engineering community with vertical-tail buffeting in the F-14 led to the development of design analysis tools and special wind-tunnel test techniques for follow-on aircraft including the F/A-18 and F-22.

F-14 model mounted in preparation for flutter tests in the 16-Foot Transonic Dynamics Tunnel.

Close-up view of the external over wing cover strakes added to prevent panel flutter.

F-14 Yaw Vanes

In the early 1980's, researchers in the Navy community became interested in the potential benefits of using thrust vectoring for control augmentation of the F-14. Their interest had been spurred by a cooperative Langley and Navy piloted simulator study that was conducted in the Langley DMS. The study defined the benefits to a representative fighter aircraft of maintaining the control power required for satisfactory V/STOL flight in conventional flight. In this study, an existing Langley simulator model of the F-14 was modified to incorporate the control modifications under the leadership of Luat Nguyen. Langley and DOD pilots flew the simulated flights.

Results of the simulator study showed that the most important benefit occurred when the yaw control was augmented at high angles of attack (normally, yaw control provided by conventional rudders is markedly reduced at high angles of attack). With the increased yaw control capability, pilots could consistently win against a variety of adversaries in simulated air-to-air combat. Analysis of the desired control levels in the simulator results indicated that deflecting the engine thrust on aircraft similar to the F-14 would provide the necessary control.

To pursue the development and demonstration of the effectiveness of yaw vectoring, the Navy conducted a series of tests to evaluate the turning effectiveness and structural integrity of external vanes mounted behind the nozzles of the F-14 engines. These activities were augmented by tests led by Bobby L. Berrier and David E. Reubush in the Langley 16-Foot Transonic Tunnel. Data obtained in these tests were used to define the geometry and thrust-vectoring effectiveness for the Navy evaluations.

Full-scale vanes were fabricated and initially ground tested behind an F-14 aircraft at the Patuxent River facility. Flight tests of a modified F-14 were subsequently conducted to demonstrate the structural integrity and thrust-vectoring performance of the vane concept over a limited flight envelope.

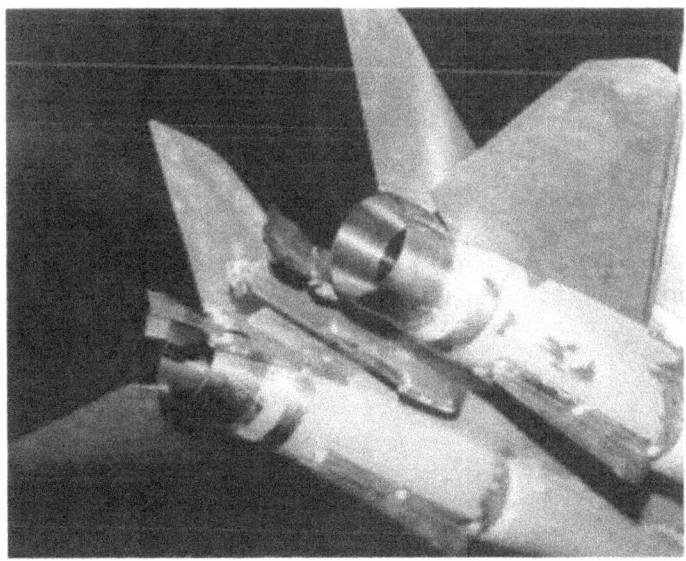

F-14 model equipped with deflectable yaw vanes in the Langley 16-Foot Transonic Tunnel.

F-14 with experimental yaw vanes in flight over Patuxent River Naval Air Station.

Further applications of the yaw vane concept to the F-14 did not develop beyond this exploratory program, but the concept of external vanes as relatively inexpensive devices for thrust vectoring at high angles of attack was successfully used in the NASA F-18 High-Angle-of-Attack Technology Program and the DARPA X-31 Enhanced Fighter Maneuverability Program.

Grumman X-29

SPECIFICATIONS

Manufacturer
  Grumman
First flight
  December 1984
Type
  Experimental demonstrator
  of advanced forward-swept
  wing and relaxed stability
Crew
  One
Engine
  General Electric F-404-GE-
  400

PARTICIPANTS

Grumman, DARPA, U.S. Air
Force, and NASA

DIMENSIONS

Wingspan . . . . 27.1 ft
Length . . . . 53.9 ft
Height . . . . 14.3 ft
Wing area . . . . 188.8 sq ft

WEIGHT

Empty . . . . 13,800 lb
Gross . . . . 17,800 lb

PERFORMANCE

Max speed . . . . . Mach number
                        of 1.6

HIGHLIGHTS OF RESEARCH BY LANGLEY FOR THE X-29

1. Langley, Grumman, Defense Advanced Research Projects Agency (DARPA), and
    U.S. Air Force cooperated to validate analytical methods for aeroelastic divergence
    of forward-swept wings.

2. Tests in the Langley 16-Foot Transonic Dynamics Tunnel of the X-29 resulted
    in advances in design methods for aeroelastically tailored forward-swept wing
    configurations.

3. Langley studied the high-angle-of-attack behavior of the X-29 and identified poten-
    tial problems such as wing rock, divergent rolling motions at post-stall angles of
    attack, and longitudinal tumbling motions, which permitted timely design of a robust
    flight control system.

4. Critical information on engine inlet performance at high angles of attack was
    obtained during tests in the Langley 14- by 22-Foot Subsonic Tunnel.

The Grumman (now Northrop Grumman) X-29 demonstrated the feasibility of several advanced technologies, including the aeroelastically tailored forward-swept wing, and the ability to routinely operate with extremely high levels of inherent longitudinal instability (relaxed static stability). Under the Defense Advanced Research Projects Agency (DARPA) sponsorship, Grumman designed two X-29 aircraft that underwent joint DARPA, Grumman, NASA, and U.S. Air Force flight tests at NASA Dryden Flight Research Center from 1984 to 1992. The exhaustive flight-test program covered aspects such as structural and aerodynamic performance, as well as high-angle-of-attack maneuverability. The X-29 aircraft flew 422 research missions. The joint X-29 Program obtained a vast amount of detailed data and analysis methods that will be applied to future high-performance aircraft.

Langley cooperated with DARPA and Grumman in the areas of flight dynamics and engine inlet performance at high angles of attack and aeroelastic divergence of forward-swept wings. Highlights of these tests included early and accurate projections of aerodynamic, stability, and control characteristics that allowed for resolution of problems before flight tests; rapid acceleration and validation of design methods for the avoidance of wing divergence for forward-swept wing configurations; and risk reduction for engine operations at high angles of attack.

Langley facilities used to support the X-29 program included the 30- by 60-Foot Tunnel, the 20-Foot Vertical Spin Tunnel, the National Transonic Facility, the 16-Foot Transonic Dynamics Tunnel, the 14- by 22-Foot Subsonic Tunnel, radio-controlled drop models, and the Differential Maneuvering Simulator.

LANGLEY CONTRIBUTIONS TO THE X-29

Evolution of Forward-Swept Wing Configurations

Research on the swept wing as a drag-reducing mechanism for high subsonic and transonic speeds during the late 1930's and early 1940's resulted in some of the first conventional aft-swept wing aircraft during World War II. At that time, it was also recognized that forward-swept wings (FSW) could produce the same beneficial effect for performance. Furthermore, the FSW also promised improved low-speed controllability. Stalls were expected to start at the wing root rather than the tip (in contrast to aft-swept wings), thereby maintaining the effectiveness of outboard ailerons and their contributions to roll control at low speeds. The onset of shock waves at high speeds was also expected to begin at the wing root, which again maintains aileron effectiveness at high speeds. With lateral control effectiveness assured across the operational envelope, there would be no need for drag-producing leading-edge high-lift devices. Finally, the general layout of the FSW resulted in a more aft location of the wing spar carry-through structure in the fuselage, which results in more fuselage internal volume.

During 1942, the German Junkers Design Bureau initiated studies of an FSW bomber designated the Ju-287. First flown in 1944, the Ju-287 exhibited several problems, the most serious of which was a tendency to increase g-loading during a turn without control inputs from the pilot. The analysis of the problem by Junkers revealed that the cause was wing structural deformation from the aerodynamic loads on the forward-facing wingtip panels. At high speeds, the deformation was predicted to become very severe, exceed structural limits, and result in wing failure. This potentially catastrophic phenomenon was referred to as aeroelastic divergence. Further analysis indicated that the structural modifications required to avoid the divergence problem for the aluminum wing of the Ju-287 would result in excessive weight and unacceptable performance penalties. Other interest in FSW configurations during World War II came from the American Cornelius Aircraft Company, which worked on several configurations, including the XFG-1, a piloted towable glider used to transport fuel.

Researchers at Langley also investigated FSW as part of a program to develop variable-sweep wings. In one of these investigations, an existing wind-tunnel model of the Bell X-1 was equipped with a variable-sweep wing, and tests were conducted for a FSW version of the aircraft in the Langley 300-MPH 7- by 10-Foot Tunnel.

Following World War II, the only significant FSW aircraft built was the German Hansajet business jet, which was designed by the same chief engineer who designed the Ju-287. The aircraft never enjoyed a large market.

The German Junkers Ju-287 forward-swept-wing bomber

The problem of aeroelastic divergence stood squarely in the progress of FSW options for relatively high-speed aircraft, and the challenge of providing sufficient rigidity versus weight turned many designers away from the concept.

In the 1970's, two activities coupled to stimulate interest in FSW configurations. First, Grumman became interested in conducting aerodynamic research to determine methods to improve its Highly Maneuverable Aircraft Technology (HiMAT) configuration, (which had lost in the design competition to Rockwell) including revolutionary wing configurations. The second activity was the remarkable advocacy and influence of Dr. Norris J. Krone, Jr., a retired Air Force pilot who had written his doctoral thesis on eliminating aeroelastic divergence of FSW configurations by using advanced tailored composites for structural rigidity. Krone subsequently became a manager at the Defense Advanced Research Projects Agency (DARPA), and Krone's discussions with the Grumman managers led to a resurgence of interest in an examination of the FSW concept. During 1977, DARPA released a request for proposals (RFP) for a highly advanced technology demonstrator that would integrate advanced aerodynamics (with emphasis on the FSW) and advanced flight controls. Responses were received from Grumman, Rockwell, and General Dynamics.

On December 22, 1981, DARPA announced that Grumman had been selected to develop the new technology demonstrator, to be known as the X-29.

High-Angle-of-Attack Technology

Initial information exchanges between DARPA and Langley on independent high-angle-of-attack evaluations of the competing FSW configurations occurred in early 1980, when Krone visited Joseph R. Chambers and his staff at the 30- by 60-Foot (Full-Scale) Tunnel. Langley agreed to provide support in this area as requested for all three competing industry teams. General Dynamics proposed an FSW version of the F-16 as a candidate design for the DARPA competition. Exploratory static wind-tunnel data had already been generated by cooperative tests led by Langley researcher Sue B. Grafton in the Langley 12-Foot Low-Speed Tunnel in 1978. Final tests of the General Dynamics design occurred in the Full-Scale Tunnel in April 1980. Rockwell's FSW configuration underwent preliminary static and dynamic tests in the Full-Scale Tunnel in March 1981. Grumman's FSW configuration was tested in the same tunnel during three entries beginning in November 1980. As a result of these tests, DARPA was provided with a timely, independent assessment of the high-angle-of-attack characteristics of all three competing designs. Langley gained considerable experience with the unique aerodynamic, stability, and control characteristics of FSW configurations at high angles of attack.

Following the award of DARPA's X-29 contract to Grumman, Langley's support in the area of high-angle-of-attack technology expanded to include additional dynamic force tests and free-flight model tests in the Full-Scale Tunnel, spin and spin recovery tests in the Langley 20-Foot Vertical Spin Tunnel, control system development studies and piloted assessments of high-angle-of-attack behavior in the Langley Differential Maneuvering Simulator (DMS), and assessments of spin-entry and post-stall motions using a radio-controlled drop model.

One of the most important, unexpected results of the X-29 high-angle-of-attack study came during preliminary static and dynamic tests of a 0.16-scale free-flight model in the Full-Scale Tunnel. The X-29 government and industry team (and the entire technical community) had expected the X-29 to be heavily damped in roll at high angles of attack as a result of the tendency of the FSW to maintain attached airflow at the wingtips during stall. However, when Sue Grafton conducted the first dynamic force tests to

Sue Grafton with the F-16 forward-swept-wing model.

measure aerodynamic damping in roll, the results indicated that the X-29 configuration would exhibit very unstable values of roll damping at angles of attack above about 25 deg. This result came as a complete surprise, and additional tests were quickly planned to confirm the suspected impact of the unstable damping. Grafton conducted a special "free-to-roll" test, in which the X-29 model was mounted to a sting assembly that contained a roll bearing which provided a 360-deg roll capability. The test technique evaluated the tendency of the model to display unsatisfactory roll characteristics at high angles of attack. At low angles of attack, the model was very stable, with no tendency to oscillate or diverge (in agreement with the results of the dynamic force test). When the angle of attack of the model was increased to about 25 deg, however, the model suddenly exhibited large amplitude wing rocking motions of a periodic nature. The wing rock was a nonlinear phenomenon, in that the model motions were self-initiating and built up to a limited amplitude independent of the magnitude of the initial disturbance.

The early identification of the wing rock led to more tests, wherein it was determined that the wings of the X-29 were indeed working as predicted. That is, the airflow remained attached at the wingtips. However, vortical flow shed from the long, pointed forebody of the X-29 was found to be interacting on the upper fuselage and inner wing and causing the unstable damping, which was so large that it overwhelmed the stabilizing influence of the attached flow at the wingtips. Interestingly, the X-29 incorporated the forward fuselage of the Northrop F-5A, which is also known to exhibit wing rock at low speeds and high angles of attack because of the same phenomenon. With the cause of the instability identified, the X-29 flight control system could be modified to increase the level of the artificial roll damping provided by feedback to the flaperons. Fortunately, the flaperons of the X-29 retained their effectiveness because of the favorable flow patterns of the FSW at high angles of attack. Estimates indicated that sufficient damping could be provided by the flight control system via the roll damper, which utilized the flaperons.

Free-flight tests of the X-29 model were first conducted in the Full-Scale Tunnel in January 1982. A special challenge faced the Langley team, since the X-29 airframe was designed for a very high level of aerodynamic instability in pitch (relaxed static stability) with a highly responsive, redundant flight control system that provided stability augmentation. The X-29 would be unflyable without the stabilizing inputs of the stability augmentation system. No other aircraft (and no other free-flight model) had ever flown with such a high level of relaxed stability. The X-29 incorporated a level of relaxed longitudinal stability (-32 percent at low speeds) that was an order of magnitude more unstable than the F-16. Under the leadership of Luat T. Nguyen, the staff programmed the X-29 control laws into the Langley computer that was used to replicate full-scale controls for the model flight tests. With a vane on the model nose boom providing information on angle of attack, the control system of the model performed flawlessly during the entire test program. The X-29 model became the first flying vehicle with such a level of relaxed stability.

X-29 free-flight model undergoing tests at high angles of attack in the Langley Full-Scale Tunnel.

During the free-flight tests led by Daniel G. Murri and Sue Grafton, the model exhibited large amplitude wing rock near an angle of attack of 25 deg when the roll damper component of the flight control system was turned off, as had been observed in the free-to-roll tests. The wing rock became more severe with increasing angle of attack, and flights usually resulted in loss of control of the model near an angle of attack of 30 deg. The amplitude and frequency of the motions were in close agreement with the preliminary tests. When the roll damper was engaged, the motions quickly damped out and the model displayed satisfactory characteristics.

The timely identification of the unstable roll damping of the X-29 configuration was a major contribution of Langley for the aircraft. If the full-scale aircraft flight program had begun without the provision for adequate control system gains and an awareness of a possible roll-damping problem, the high-angle-of-attack characteristics of the X-29 could have been unacceptable. Ultimately, the X-29 did display the wing rock tendency in flight (although not to the degree of severity indicated by the model tests). However, the controls of the full-scale aircraft were even more effective than those of the free-flight model, and the control system had been designed with provision to increase the gain of the roll damper. The two X-29 aircraft subsequently exhibited entirely satisfactory characteristics in flight, and the technical community learned a lesson regarding the integrated aerodynamic contributions of FSW configurations with long, pointed fuselages.

In 1983, Luat Nguyen led a piloted assessment of the X-29 at high angles of attack in the DMS. The objectives were to study the high-angle-of-attack flying characteristics of the aircraft and its susceptibility to departure under maneuvering conditions, identify and develop control law concepts to provide good flying qualities and a high level of departure-spin resistance, assess the effectiveness of the Grumman flight control laws at high angles of attack and provide recommendations for modifications, and provide support for flight-test planning and coordination with NASA Dryden Flight Research Center.

A priority issue in the DMS studies was the effect of Reynolds number on aerodynamic characteristics. Results of high Reynolds number tests in the NASA Ames 12-Foot Pressure Tunnel indicated that significant effects of Reynolds number on pitching moments and yawing moments existed for the X-29 configuration. Working with flight control team members from Grumman and Dryden, Nguyen completed an in-depth study of X-29 handling qualities at high angles of attack and provided numerous recommendations for the specific design of the flight control system for high-angle-of-attack conditions. Control law concepts were identified for control surface interconnects for optimum roll coordination, wing rock suppression, and automatic departure and spin prevention.

Spin Tunnel Tests

Tests of the X-29 in the Langley 20-Foot Vertical Spin Tunnel, which began in 1981 under Raymond D. Whipple, concentrated on three areas. First, the developed spin and spin recovery characteristics of the X-29 were determined for various aircraft loadings and erect and inverted spins. As previously mentioned, large Reynolds number effects had been predicted for the X-29 configuration. These effects were very noticeable at very high angles of attack (near 90 deg), where the model might exhibit critical spins. To correct for these effects (which were caused by the unique forebody shape that was incorporated from the F-5A) at the low speeds involved in the Spin Tunnel conditions, the Langley staff employed auxiliary strakes on the nose of the X-29 spin model. The

X-29 model exhibited two types of spin. One type was flat, with an angle of attack of about 85 deg, with marginal to unsatisfactory spin recovery. The second spin was very oscillatory, with satisfactory to excellent recovery characteristics.

The second area of interest was determination of the size of the emergency spin recovery parachute required for the number 2 aircraft, which would be used in the high-angle-of-attack flight-test program. Working with Grumman and Dryden, Whipple and the X-29 team arrived at a recommended parachute size, truss structure, and deployment mechanism. As a result of the outstanding maneuverability and spin resistance exhibited by the X-29, the emergency parachute system was never utilized in flight tests to terminate post-stall maneuvers.

The third area of interest in the Spin Tunnel test program involved concern over the possibility of a longitudinal tumbling phenomenon for the X-29. This concern had risen because of the high level of inherent longitudinal instability designed into the X-29. Specifically, Langley researchers expressed concern over whether the combination of very low airspeed and very high angle of attack (such as during recovery from a "zoom climb" to zero airspeed) might result in the aircraft pitching over into an end-over-end tumbling that would result in incapacitating g-levels for the pilot. The Spin Tunnel staff first addressed this issue in free-spinning tests wherein the X-29 model was launched tail first, without rotation, into the vertically rising airstream. The results of these exploratory tests showed that, without inputs from the control system, the model would indeed pitch over and develop a continuous tumbling motion about its center of gravity. The wild gyrations quickly caused the spin tunnel model to impact the walls of the tunnel, so an additional test technique was developed to study the issue under more controlled conditions. In these tests, the model was mounted on a special single-degree-of-freedom test apparatus that permitted 360-deg free-pitching motions. The additional

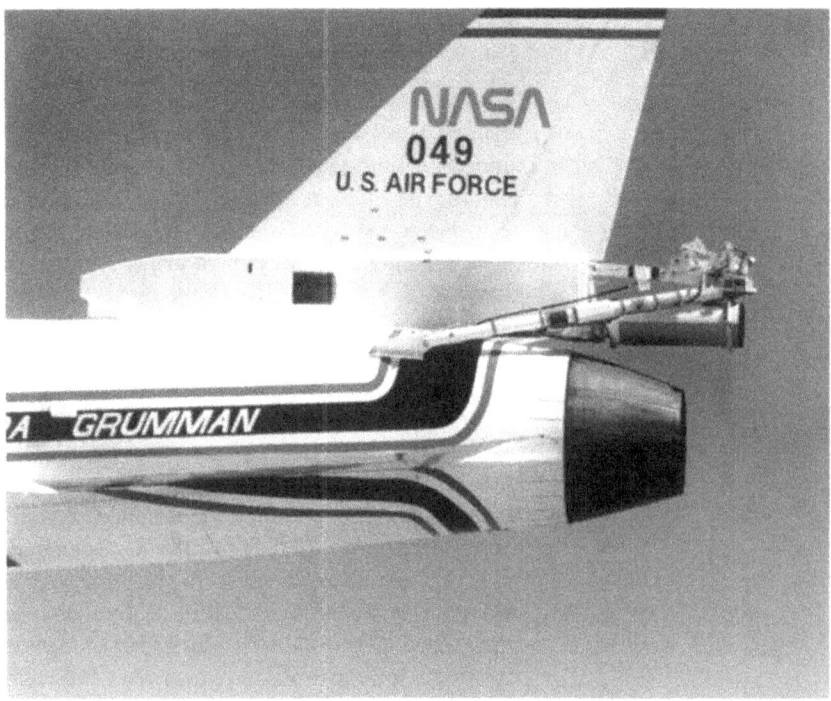

The spin parachute and truss assembly on the X-29 during high-angle-of-attack tests at Dryden.

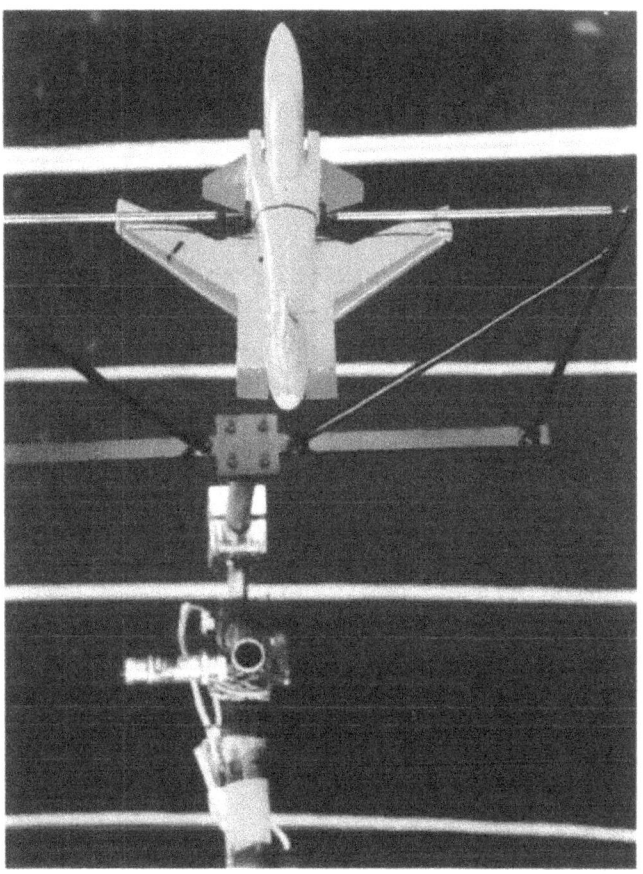

X-29 model mounted on single-degree-of-freedom
pitch-tumble test apparatus in the Langley Spin Tunnel

studies of tumbling on this apparatus provided insight to a possible solution to the problem. The X-29 flight control system included aft-fuselage strake flaps, which were intended to be used only as trimming devices. However, during the tumbling tests it was found that the strake flaps were extremely efficient in promoting recovery from the tumble motions. In view of this result, the control system was modified to permit use of the strake flaps as control devices. This approach prevented the X-29 from entering the uncontrollable tumbling motions.

Drop-Model Tests

In 1987, Langley upgraded its drop-model test technique and conducted studies of the high-angle-of-attack and post-stall characteristics of a 0.22-scale model of the X-29. The high degree of relaxed stability of the X-29 made high-fidelity simulation of the control system a mandatory feature of the drop model. Langley had never flown an unstable model before the X-29 program. The Langley staff, led by Mark A. Croom, upgraded nearly every element of the drop-model operation, such as the model control actuators, transmitters and receivers, data encoders and decoders, and operational displays including a new cockpit display and operations monitor. The advanced X-29 control laws developed in the DMS study were programmed into a ground-based computer for proper simulation of control effects. This activity was by far the most challenging drop-model project ever conducted by Langley to that time.

The results of the drop-model program further confirmed the need for special control system concepts for high-angle-of-attack conditions. Without artificial stability augmentation in roll, the drop model exhibited the same type of large amplitude wing rock motions previously displayed by the wind-tunnel free-flight model. In the case of the drop model, when the angle of attack was increased to 30 deg and beyond, the oscillations became so divergent that the model exhibited uncontrollable 360-deg rolls that evolved into a roll departure and post-stall gyrations. When the wing rock suppression system was engaged, the motions were damped and the model was controllable to very high angles of attack.

During some flights at extreme angles of attack, asymmetric yawing moments were encountered that caused the model to yaw and generate relatively high rotational rates. When the pilot intentionally applied prospin control inputs during these conditions, the model sometimes entered the unrecoverable flat spin mode exhibited by the spin tunnel model. With the departure and spin prevention system engaged, the model was highly resistant to intentional spins and was extremely maneuverable at high angles of attack.

Another valuable contribution of the X-29 drop-model program was a parameter identification effort conducted by David L. Raney and James G. Batterson of Langley, in which they analyzed the wing rock motions of the drop model and extracted values of the critical aerodynamic parameters that caused the motions. This complex analysis required the identification of rapidly changing aerodynamic derivatives over a range of angle of attack. The results of the study clearly identified unstable values of aerodynamic damping in roll to be the cause of the wing rock motions.

X-29 drop model mounted on launch apparatus on side of helicopter prior to release for post-stall tests.

Engine Inlet Tests

Grumman requested support from Langley in assessing the static and dynamic pressure conditions that might be experienced by the engine inlets during the high-angle-of-attack flight tests of the X-29. The data were required to ensure satisfactory engine operation and avoid engine stalls that could result in loss of hydraulic power during the spin maneuvers. Langley responded with tests of a powered X-29 model in the 14- by 22-Foot Subsonic Tunnel. The model was equipped with a propulsion simulator, flow-through inlets, and extensive instrumentation. These tests occurred in 1991 under the leadership of John W. Paulson, Jr.

X-29 powered-inlet model tests in the Langley 14- by 22-Foot Subsonic Tunnel.

Aeroelastic Divergence

The phenomenon of aeroelastic divergence had dramatically constrained international interest in the application of conventional metal FSW concepts. Franklin W. Diederich and Bernard Budiansky of Langley studied and summarized the major challenges of the divergence phenomenon in a NACA Technical Note in 1948 (ref. 1). However, the emergence of composite materials and the aggressive advocacy of Norris Krone to utilize aeroelastic tailoring to reduce divergence led to cooperative studies of FSW technology by the staff of the Langley 16-Foot Transonic Dynamics Tunnel (TDT). Rodney H. Ricketts, Robert V. Doggett, and Wilmer H. Reed, Jr. planned and participated in numerous studies with DARPA, the Air Force, and industry to develop and verify analytical predictions of the divergence phenomenon. The studies investigated systematic generic wings and specific configurations, including the Rockwell and Grumman FSW designs.

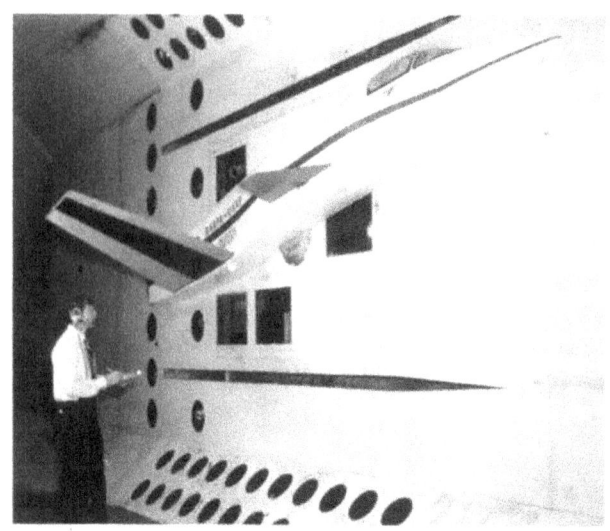

Rodney Ricketts with the Grumman FSW model during aeroelastic divergence tests in the Langley 16-Foot Transonic Dynamics Tunnel.

The analyses and tests of generic wing models with variations in aspect ratio, airfoils, and sweep provided invaluable data and methods that significantly expanded the database, especially in the transonic regime. Six subcritical response test techniques were formulated and evaluated at transonic speeds for accuracy in predicting static divergence, and two divergence stoppers were developed and evaluated for use in preventing structural damage of wind-tunnel models during divergence tests.

Ricketts led cooperative tests of the proposed Grumman and Rockwell FSW configurations in the TDT in 1979. NASA, DARPA, the U.S. Air Force, and the industry teams participated in the cooperative tests. The Rockwell model consisted of a semispan wing mounted to a splitter plate on the tunnel sidewall. The Grumman model included a representative fuselage shape. Conclusions from the divergence tests included dramatic demonstrations that aeroelastic tailoring was extremely effective in suppressing divergence for FSW configurations and that new nonlinear aerodynamic theories were required for complete analysis of the phenomenon. After the DARPA X-29 contract was awarded to Grumman, additional tests were conducted in the TDT in 1983 which demonstrated the potential coupling of the wing structural modes with the rigid body pitch mode to create an instability called body-freedom flutter. Techniques were developed to analyze these wing-body interactions, and the wing divergence prediction methods developed at Langley were used in the flight-test program of the X-29 at Dryden.

Advanced Engine Nozzle

In the early 1970's, an advanced thrust-vectoring nozzle for V/STOL aircraft was designed by the General Electric Company under a contract funded by the Navy. The nozzle was designed with a deflectable internal hood to permit large pitch thrust-vector angles for V/STOL operations, while deflection of the single expansion ramp could be used for smaller thrust-vector angles during air-to-air combat. This nozzle, known as the augmented deflector exhaust nozzle (ADEN), received considerable test and developmental support in the technical community during the 1970's and early 1980's. Much of this activity was advocated by the Department of Defense (DOD), NASA, the U.S. Air

Simple ADEN-like nozzle configuration (pitch and yaw vectoring) flown on X-29 free-flight model in the Full-Scale Tunnel.

Force, and the U.S. Navy ad hoc interagency nonaxisymmetric nozzle working group that included Langley researcher Bobby L. Berrier as the NASA representative. The Propulsion Aerodynamics Branch at Langley conducted much of the development work on the ADEN under the direction of William P. Henderson and Bobby Berrier. Numerous cooperative ADEN research programs with DOD and industry were conducted to optimize nozzle performance and define suitable propulsion-airframe integration methodologies on a series of generic and specific (e.g., F-18) fighter configurations. The first full-scale nonaxisymmetric nozzle test was conduced at the Glenn Research Center in 1976 by running an ADEN nozzle attached to an augmented General Electric YJ101 engine in an altitude test cell. Thus, by the time of the DARPA X-29 award to Grumman in 1981, a mature thrust-vectoring nozzle design was available.

During the early phases of the X-29 program, NASA Headquarters expressed interest in the application of the ADEN to the canard configured X-29 as an exploratory flight-test bed, since the canard of the X-29 could counterbalance the effects of the deflected ADEN nozzle. The NASA interest in the X-29 ADEN was to promote technical progress in vectorable nozzles and ensure that a sufficient number of candidate nozzles were considered for further development.

NASA Headquarters requested Langley's assessment of the X-29 ADEN configuration for further advocacy discussions with Grumman and DARPA. Although no induced lift or aerodynamic wing lift augmentation would occur for this configuration, the addition of vectoring was believed by Langley to improve high-angle-of-attack controllability. In response to this request from Headquarters, Joseph L. Johnson, Jr. and his staff at the Full-Scale Tunnel quickly modified their X-29 free-flight model and conducted exploratory evaluations of the handling qualities of the X-29 ADEN at high angles of attack in 1984. During these tests, a yaw-vectoring side-door capability (a concept defined and tested by the Propulsion Aerodynamics Branch in the Langley Jet Exit Test Facility) was added to a simplified ADEN-type nozzle to provide both pitch and yaw vectoring, and the resulting flight tests demonstrated a dramatic improvement in controllability and agility of the X-29 at extreme angles of attack. The model could be flown with precise and effective control to angles of attack as high as 85 deg. These positive results were typical of those that had been obtained with several different aircraft configurations equipped with other thrust-vectoring nozzles. The national progress in nonaxisymmetric, thrust-vectoring nozzles led to interest in other nozzle configurations and the eventual successful F-15 short takeoff and landing and maneuver technology demonstrator (STOL/MTD), F-18 High Alpha Research Vehicle (HARV), and X-31 flight-test programs.

X-29 ADEN model flying at an angle of attack of 80 deg in the Langley Full-Scale Tunnel.

Lockheed C-5 Galaxy

SPECIFICATIONS

Manufacturer
  Lockheed
Date in service
  C-5A . . . . . . December 1969
  C-5B . . . . January 1986
Type
  Transport
Crew
  Six
Engine
  General Electric TF39-GE-1C

USERS

U.S. Air Force (Active, Air National Guard, and Reserve)

DIMENSIONS

Wingspan . . . . 222.8 ft
Length . . . . 247.8 ft
Height . . . . 65.1 ft
Wing area . . . . 6,200 sq ft

WEIGHT

Empty . . . . 370,000 lb
Max takeoff . . . . 837,000 lb

PERFORMANCE

Cruise speed . . . . 563 mph
Range . . . . 3,000 n mi

HIGHLIGHTS OF RESEARCH BY LANGLEY FOR THE C-5

1. The aerodynamic performance of three competing industry configurations for the C-5 contract was determined by model tests in the Langley 8-Foot Transonic Tunnel.

2. Aerodynamic interference between the wing and the large engine nacelles and pylons were measured during tests in the 8-Foot Transonic Tunnel.

3. Parametric tests in the Langley 16-Foot Transonic Dynamics Tunnel of a clipped-span C-5 model identified an abrupt drop in tail flutter speed at transonic Mach numbers that required increased stiffening of the fin spar.

4. An active load alleviation system was developed in the Langley 16-Foot Transonic Dynamics Tunnel.

5. Powered-model tests in the Langley 30- by 60-Foot (Full-Scale) Tunnel provided information on C-5 airdrop (wake) characteristics and power effects for the landing configuration.

6. Tests of a C-5 model were conducted to determine an optimum ditching configuration.

The Lockheed (now Lockheed Martin) C-5 Galaxy heavy-cargo transport provides strategic airlift for the worldwide deployment and supply of combat and support forces. The C-5 can carry unusually large and heavy cargo for intercontinental ranges at high subsonic speeds. The aircraft can take off and land in relatively short distances and taxi on substandard surfaces during emergency operations. The C-5 is one of the largest aircraft in the world. It is almost as long as a football field and is as tall as a 6-story building with a cargo compartment about the size of an 8-lane bowling alley. The C-5 and the smaller C-141B Starlifter are strategic airlift partners. Together they can carry fully equipped, combat ready troops to any area in the world on short notice and provide the full field support necessary to maintain a fighting force. The C-5 can carry a payload that is more than twice as heavy as the C-141 payload. The lower deck of the C-5 has an unobstructed length of 121 ft and width of 19 ft that enables it to carry any piece of Army equipment, including self-propelled howitzers, personnel carriers, and tanks—none of which can enter the payload bay of the C-141.

Langley's contributions to the C-5 program included wind-tunnel tests and performance analysis of the competing industry designs during the Cargo Experimental–Heavy Logistics System (CX–HLS) Program that resulted in the C-5. Langley conducted wind-tunnel assessments of engine and wing aerodynamic interactions; flutter studies of the T-tail configuration; flutter clearance tests for the complete configuration; wind-tunnel studies of an active load alleviation concept; flow surveys behind the configuration for analysis of airdrop characteristics; assessments of power-induced effects in the landing configuration, including studies of potential applications of the externally blown flap concept (subsequently incorporated in the C-17 transport); and model tests to determine the optimum ditching configuration. Langley facilities involved in C-5 tests included the 8-Foot Transonic Tunnel, the 16-Foot Transonic Dynamics Tunnel (TDT), the 16-Foot Transonic Tunnel, the 30- by 60-Foot (Full-Scale) Tunnel, and the Langley Impacting Structures Facility (Tow Tank).

LANGLEY CONTRIBUTIONS TO THE C-5

C-5 Evaluation

In 1964, the Air Force awarded study contracts to Boeing, Douglas, and Lockheed for the design of a heavy-lift transport for the Cargo Experimental–Heavy Logistics System (CX–HLS). In December 1964, the three manufacturers were invited to submit proposals for the new transport, which was intended to carry bulky and heavy military equipment that could not be accommodated in the C-141. All three industry designs incorporated high-wing configurations with four large turbofan engines in underwing nacelles and front and rear doors with ramps for flow-through loading and unloading. The Boeing and Douglas designs had conventional tail configurations, whereas the Lockheed design incorporated a T-tail configuration.

At the request of the Department of Defense (DOD), Langley conducted aerodynamic assessments of all three designs in the Langley 8-Foot Transonic Pressure Tunnel in 1965 under the leadership of Langley researcher Dr. Richard T. Whitcomb. The results of these tests were provided to a committee to select the winning C-5 design. In an extremely controversial decision based on proposal cost estimates, industry workloads, and geopolitical considerations, the Air Force announced in October 1965 that Lockheed had been selected to proceed with development of their C-5 design.

The C-5 design submitted by Boeing was found to have superior aerodynamic cruise performance in the transonic wind-tunnel tests performed at Langley. Boeing's experience with the C-5 competition coupled with Boeing management's vision of the marketability of jumbo civil transports (and interest from Pan American Airlines) led to the development of the Boeing 747, which enabled Boeing to dominate the world market with a new product line. Although the 747 was a completely new aircraft design (low wing, passenger-carrying civil aircraft), the general configuration influence of the earlier C-5 candidate is in evidence.

Research on Sting Interference Effects

Analysis of the data from the Langley 8-Foot Transonic Pressure Tunnel tests of the three C-5 configurations indicated that the high degree of aft-fuselage bottom upsweep of all the configurations increased cruise drag (especially the Douglas design, which had 19 deg of upsweep). Because of the critical nature of the aft-end drag, concern arose over potential interference effects caused by the model support sting on drag measurements.

Donald L. Loving and Arvo A. Luoma conducted tests in 1965 of all three configurations with conventional stings, dummy stings, and dorsal strut-support systems over a limited range of test variables from just below to just above the design cruise condition of each configuration. The results of their investigation indicated that the sting interference effects were of very small magnitude.

Research on Effects of Test Section Size

Also in 1965, a study was conducted by Langley researchers Arvo Luoma, Richard J. Re, and Donald Loving to address concerns over the effect of model size during transonic tests of large models in the 8-Foot Transonic Pressure Tunnel. In wind-tunnel tests, the largest model possible is generally desirable so that higher model Reynolds numbers are obtained, but if the model is too large potential tunnel wall effects and data corrections become concerns, especially at high subsonic and transonic speeds. Comparative aerodynamic data were obtained for the same 5-ft-span model of the C-5 in the

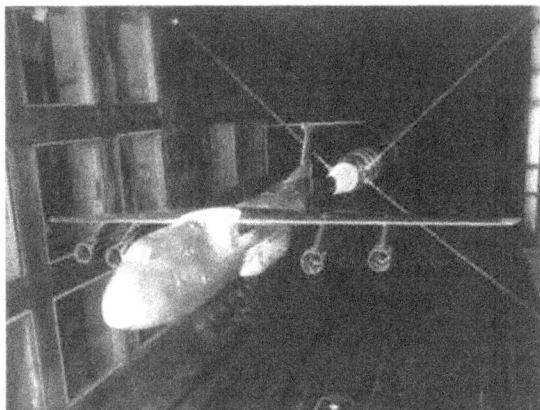

Competing C-5 configurations during tests in the Langley 8-Foot Transonic Pressure Tunnel. Top to bottom: Douglas, Lockheed, and Boeing designs.

C-5 model mounted for tunnel test section study in the Langley 16-Foot Transonic Tunnel

Langley 8-Foot Transonic Pressure Tunnel and the Langley 16-Foot Transonic Tunnel for Mach numbers from 0.75 to 0.83. The 5-ft-span model was two to three times larger than usual models tested in the Langley 8-Foot Transonic Pressure Tunnel.

The results of the study indicated that the data obtained in the two tunnels were in good agreement and that large models could be tested in a slotted tunnel such as the 8-Foot Transonic Pressure Tunnel at subsonic speeds with acceptable results.

Propulsion Integration Research

In 1966, Langley researchers James C. Patterson, Jr. and Stuart G. Flechner conducted parametric studies in the Langley 8-Foot Transonic Pressure Tunnel to determine the effects of large high-bypass engines on the interference drag of wing-nacelle configurations at cruise conditions. In the study, a large powered semispan model that incorporated tip-driven, nitrogen-powered engine simulators was used. The baseline model configuration represented the C-5; however, the horizontal and vertical positions of the engines relative to the wing were varied so the effects of the engine exhaust wakes could be studied in detail. The results of the study indicated that the interference drag effects could actually be beneficial for certain combinations of wing-nacelle geometric parameters.

Powered C-5 semispan model in the Langley 8-Foot Transonic Pressure Tunnel.

Low-Speed and Wake Characteristics

In late 1965, the Air Force requested that tests be conducted in the Langley 30- by 60-Foot (Full-Scale) Tunnel to investigate the application of the externally blown flap to the C-5 aircraft. Close examination of the C-5 configuration indicated that the engine exhaust would blow strongly on the trailing-edge flaps and that the aircraft would probably have considerable jet-flap lift effect in its basic configuration. The jet-flap lift effect might be sufficient to promote marked improvements in takeoff and landing performance that had not been anticipated. However, the jet-flap effect could induce pitch and roll trim problems that were also not anticipated. The Air Force request included tests of a semispan powered model, as well as a full-span powered model.

The examination by Langley of the C-5 configuration indicated that several features of the wing and flap did not lend themselves well to adaptation of a high efficiency jet-flap arrangement. For example, the flap configuration was not optimum for blown flap applications. As a result of concern expressed by Langley, a two phase program was conducted. The first phase of the program measured the effects of power on lift and drag characteristics and the effect of power (and engine-out conditions) on the longitudinal and lateral trim requirements, control, and stability of the basic C-5 configuration. The second phase of the research program consisted of modifying the wing and flap system to provide a more optimal configuration for an efficient jet flap.

In 1966, a powered semispan model of the C-5 underwent two test entries in the Full-Scale Tunnel to determine a limited amount of power effects, including the effects of thrust reversers. Unfortunately, Lockheed changed the trailing-edge flap system of the C-5 from a double-slotted configuration to a Fowler flap configuration, and the tests had to be conducted with an outdated flap system.

In 1967, a full-span 0.057-scale model of the C-5 was tested in the Full-Scale Tunnel to determine the effect of the externally blown flap concept on the aircraft. The initial work was done with the original, nonoptimized double-slotted flap design. Subsequent work was done with a new flap system that was designed specifically for the blown flap concept by the Langley staff under the leadership of Joseph L. Johnson, Jr.

Powered C-5 model mounted in the Langley Full-Scale Tunnel for externally blown flap assessments with the double-slotted flap configuration.

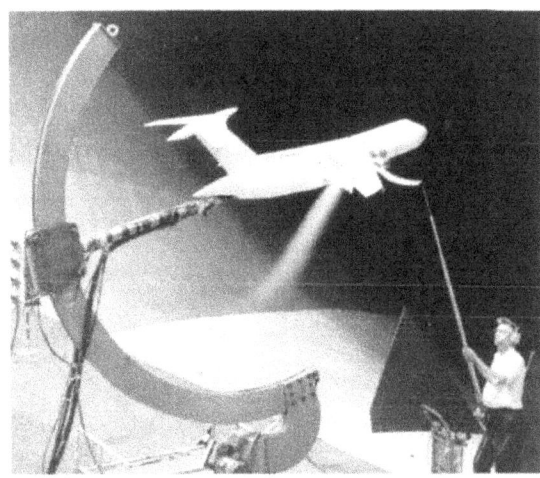

Langley engineer Charles C. Smith, Jr. conducts flow visualization studies on the modified C-5 model.

The results of the wind-tunnel study indicated that, as had been predicted by Langley, the blown flap effectiveness for the original double-slotted flap configuration was relatively modest. Modifications required to obtain a more optimal externally blown flap configuration were fairly extensive, including new trailing-edge flaps, relocation of the horizontal tail, and resizing of the vertical tail and rudder for control of engine-out conditions. Although such modifications were beyond the scope and interests of the C-5 program, Langley conducted additional wind-tunnel tests of the modified C-5 configuration, including free-flight model tests and piloted simulator studies. Although not applied to the C-5, this research provided critical information, design guidelines, and solutions to engine-out issues that were extremely helpful in the risk reduction and ultimate development of the C-17 transport.

At the request of the Army, a survey of the flow in the wake of the C-5 full-span model was made in early 1968 in the Full-Scale Tunnel to define the dynamic pressure and flow angularity for analyses of the airdrop capability of the C-5.

Tail Flutter

The staff of the Langley 16-Foot Transonic Dynamics Tunnel (TDT) led research on flutter characteristics of the T-tail configuration as a design feature of large transport aircraft with transonic cruise capabilities in the late 1950's. Although several aerodynamic theories had been developed for predicting subsonic and supersonic flutter of T-tails, the transonic regime posed special challenges because transonic flutter speeds tend to be lowest and are accompanied by complex shock patterns that make flutter analyses difficult. TDT staff members Charles L. Ruhlin and Maynard C. Sandford had been actively involved in the flutter issues that faced the C-141 and had contributed to the databases for the design of flutter-free T-tails for future aircraft. Their pioneering efforts in the C-141 program resulted in new test procedures, validation of computational methods, and fundamental research of the complex aerodynamic and structural coupling for the new tail designs. Two types of T-tail flutter had been identified by research: symmetric tail flutter, which was dominated by pitching-type relative motions of the surfaces, and antisymmetric flutter, which was characterized by yawing and rolling relative motions of the tail surfaces. Antisymmetric flutter is especially complex, and more data were needed to advance the understanding and design tools.

Clipped wing model of the C-5 in the Langley 16-Foot Transonic Tunnel for flutter tests.

Full-span model of the C-5 in the Langley 16-Foot Transonic Dynamics Tunnel.

Ruhlin and Sandford developed an innovative approach to testing large, isolated-tail and aft-fuselage models, which permitted more accurate simulation of structural and aerodynamic properties of full-scale aircraft. In late summer of 1966, they began studies of an isolated 1/13-scale model of the C-5 empennage, fuselage, and inner wings in the TDT. The T-tail, fuselage, and inboard wings were geometrically, dynamically, and elastically scaled. Two models of the T-tail were tested—one was built with the C-5 design stiffness, while the second had only half the design stiffness.

The tail flutter speeds for the designed C-5 empennage were beyond the required flutter demonstration speeds; however, a pronounced decrease in flutter speed was observed slightly above the maximum Mach number (between 0.92 and 0.98). As a result of these tests, the fin spar stiffness was increased, and flutter clearance for the full-span C-5 model in the TDT was subsequently obtained.

Load Alleviation

In the initial design of the C-5, Lockheed implemented an aggressive weight reduction program to meet performance requirements. The wing weight was reduced by using higher design stress levels and reducing primary component thickness. The higher stress levels proved to be a problem, and wing cracks were found early in full-scale ground fatigue tests in July 1969. After the aircraft had been in service several years, a wing tear-down inspection on one aircraft with a high number of flight hours revealed significant cracks. Lockheed proposed several approaches to restore C-5 wing fatigue life to a specified level of 30,000 flying hours. These approaches included (1) an active aileron system to alleviate gust loads on the wing, (2) local wing modifications to improve fatigue, and (3) redistribution of fuel within the wing to reduce bending moments.

Langley researchers Ruhlin and Sandford conducted tests of the C-5 active lift distribution control system (ALDCS) in the 16-Foot Transonic Dynamics Tunnel in 1973. The results of this study validated the use of active control technology for the minimization of aircraft aeroelastic response and showed that scaled aeroelastic wind tunnel models can be used in developing active control systems.

Active ailerons were retrofitted to 77 C-5's in 1975 through 1977. That approach, however, was superseded by a redesign of the wing that included a new center wing, two inner wing boxes, and two outer wing box sections, which were manufactured from advanced aluminum alloys that were unavailable when the original wings were produced in the late 1960's and early 1970's. As a result, all C-5 aircraft were modified with the new wings.

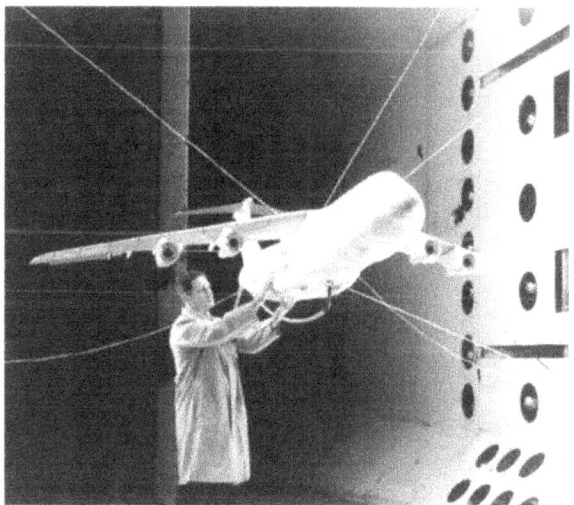

Active load alleviation test of the C-5 in the
Langley 16-Foot Transonic Dynamics Tunnel.

Ditching Tests

Langley conducted ditching investigations for military and civil aircraft (including the Space Shuttle) for many years in a water tank facility known as the Langley Impacting Structures Facility. Following World War II, aircraft shapes and sizes did not vary significantly from the existing database that was generated by the Langley research. However, the introduction of the C-5 and other large wide-body civil transports required the prediction of ditching characteristics for heavier and larger configurations that were not

C-5 model in Langley Impacting Structures Facility for ditching tests.

C-5 ditching model with simulated structural skin on bottom of model.

included in the database. The design configuration and structural features of large cargo and transport aircraft also required that a dynamic model be investigated to determine overall motions, accelerations, and the approximate location and amount of damage that might be expected during a ditching at sea. In addition, the large number of main landing gear wheels (24) for the C-5 and the ability of the landing gear to be extended to various positions offered the possibility of an optimal ditching configuration, since previous wheels-down dynamic-model investigations had shown a wide variation of ditching performance with landing gear extended.

At the request of the Air Force, Langley researcher William C. Thompson conducted ditching studies of a 1/30-scale model of the C-5 in the Impacting Structures Facility in 1969. The model was highly instrumented to measure accelerations, and a special scaled-structure lower fuselage was included to determine the structural damage incurred in ditching. The tests included various impact attitudes, flaps up and down, and landing gear retracted and down in simulated calm and rough (simulated sea state 4) water. Results of the study indicated that the most favorable condition for ditching was a 7-deg nose-high attitude with the flaps down 40 deg, the nose gear retracted, and the main gear fully extended. For these conditions, damage to the fuselage bottom would occur, and most of the main landing gear would probably be torn away.

Wind-Tunnel Test to Flight-Test Correlation

Analysis of flight data obtained with the C-5 indicated that significant differences existed between pressure measurements made in the wind tunnel and flight. Similar issues had previously arisen during the C-141 program. These issues involved difficulties with scaling effects in wind-tunnel tests. Extensive fundamental research was initiated in the Langley 8-Foot Transonic Pressure Tunnel under the direction of Richard Whitcomb. The research was initiated to develop an approach of simulating high Reynolds number flows by using specific placement of artificial boundary-layer trip strips on the wing upper and lower surfaces of the wind-tunnel model. The test program, which was conducted by Donald Loving, Arvo Luoma, and James A. Blackwell, Jr., successfully identified a methodology that more properly simulated the appropriate boundary-layer thickness in transonic wind-tunnel tests. This technique was used extensively in the development of the supercritical airfoil.

Lockheed Martin C-130 Hercules

SPECIFICATIONS

Manufacturer
　Lockheed Martin
Date in service
　C-130A 1956
　C-130E 1962
　C-130J 1996
Type
　Transport
Crew
　Three
Engine
　Allison T56-A-15 turboprops

USERS

U.S. Air Force (Active, Air
National Guard, and Reserve),
U.S. Navy, U.S. Marine Corps,
U.S. Coast Guard, and over
65 countries

DIMENSIONS

Wingspan . . . . 132.6 ft
Length . . . . 97.8 ft
Height . . . . 38.4 ft
Wing area . . . . 1,745 sq ft

WEIGHT

Empty . . . . 75,743 lb
Max takeoff . . . . 175,000 lb

PERFORMANCE

Cruise speed . . . . .360 mph
Range . . . . .2,700 n mi

HIGHLIGHTS OF RESEARCH BY LANGLEY FOR THE C-130

1. Lockheed used aerodynamic computational codes developed by Langley in the configuration development of models of the C-130.

2. Under a Langley contract in the early 1970's, Lockheed produced an advanced boron reinforced metal center wing box for the C-130 that was flight tested on three C-130E aircraft.

3. In the 1980's, Lockheed tested a composite center wing box for the C-130, which was not implemented but provided expertise for applications of advanced composites to other Lockheed Martin products, including the F-22.

4. Lockheed, under contract to Langley, developed piloted simulators and interacted with Langley researchers to develop advanced cockpit liquid crystal flat-panel displays for the C-130J.

The Lockheed (now Lockheed Martin) C-130 Hercules is arguably the most versatile military transport aircraft ever built. The aircraft is capable of operating from rough, dirt strips and is the prime transport for air dropping troops and equipment into hostile areas. C-130's fulfill a wide range of operational missions in both peace and war, with specialized versions of the aircraft that perform an enormous number of roles. The C-130J incorporates state-of-the-art technology to reduce manpower requirements, lower operating and support costs, and provide life cycle cost savings over earlier C-130 models.

Langley's contributions to the C-130 program include composite structures and materials, noise reduction, cockpit displays, and aerodynamic technologies. Lockheed has extensively used basic and applied research studies conducted by Langley in the continuous evolution of the C-130. For example, aerodynamic analyses of wing performance, stall characteristics, and loads have been accomplished with computational methods developed by Langley. Also, Langley methodology has been used to predict C-130 exterior noise and to reduce interior noise. Lockheed's participation in the design and development of advanced simulators at Langley and Langley's research on advanced cockpit displays led to the adoption of liquid crystal flat-panel displays for the C-130J. The most interactive Langley and Lockheed studies of the C-130 have occurred in applications of composite structures and materials. Under Langley sponsored contracts in the 1970's, Lockheed fabricated and tested a boron reinforced metal center wing box to improve the fatigue life by a factor of three and reduce the weight by 300 lb. The modified wing box was subsequently flight tested on three C-130 aircraft, which are still in routine fleet service. In the 1980's, a subscale all-composite C-130 center wing box was designed, fabricated, and structurally tested as part of the Langley Advanced Composites Technology (ACT) Program to validate the potential for increased load levels with reduced weight.

On November 6, 1998, Lockheed Martin arranged for a C-130J to visit Langley for a special ceremony in honor of NASA's continuous contributions to the C-130 Hercules.

LANGLEY CONTRIBUTIONS TO THE C-130

Composite Structures and Materials

Lockheed Martin has had an interest in Langley research on the development of advanced composite structures for the past 30 years. Working under a long series of Langley contracts, Lockheed has developed composite flaps, wing trailing edges, and propellers for the C-130J. As a result of extensive experience from these efforts, Lockheed has also developed composite process improvements for reduced costs.

In 1968, thorough inspections of the Air Force C-130 fleet revealed that almost half of the 619 aircraft in operational service had fatigue cracks in the center wing section. The technical community offered numerous recommendations to update and repair this vital aircraft.

In the 1970's, Lockheed conducted a five-phase program under a Langley contract to develop and demonstrate the selective reinforcement of conventional metallic structures with boron composites. This program, under the direction of Langley researcher H. Benson Dexter, was designed to demonstrate that advanced filamentary boron composites could improve static strength and fatigue endurance with less weight than would be possible with metal reinforcement. Contract activities included the development of a basis for structural design, selection, and verification of materials and processes; manufacturing and tooling development; and fabrication and test of full-scale portions of the center wing box. The baseline C-130E aluminum center wing box design was modified by removing aluminum and adding unidirectional boron reinforcing laminates bonded to the crown of the hat stiffeners in the wing structure. The center wing sections on three C-130E fleet aircraft with fatigue cracks were replaced with the new stronger aluminum wing box. This hybrid design improved the fatigue life by a factor of three while reducing the weight by 300 lb over the conventional metallic wing. Flight tests of the modified center wing structure were later conducted on three C-130E fleet aircraft. These aircraft were entered into routine fleet service in 1974 and are still operational. Unfortunately, the Air Force was not ready to commit to the new technology.

In the 1980's, Lockheed actively participated in several contracts for the development of composite wing and fuselage structures that led to the Langley Advanced Composites Technology (ACT) Program. Working under a Langley contract managed by Randall C. Davis, Lockheed designed and fabricated a full height, half-chord, all-composite center wing box for the C-130. This wing box was designed, fabricated, and successfully tested at the higher load requirements of the special operations aircraft. The resulting wing box weight was lower than the conventional metal wing box. Funding reductions for the ACT Program terminated the potential C-130 applications, and the remaining advanced composite program elements were applied to other aircraft such as the F-22.

Advanced Cockpit Displays

Under NASA contract, Lockheed built three individual simulators that are located at Langley, NASA Ames Research Center, and Lockheed. Interactions with the Langley staff regarding NASA research on advanced cockpit displays led to the development of head up display (HUD) technology and liquid crystal flat-panel displays, which provide integrated instrument information and significantly reduce pilot workload and increase maneuver precision for the C-130J.

Acoustics

As a result of certain mission specific requirements, the external and interior noise characteristics of the C-130 have received considerable attention from Lockheed. The utilization of the aircraft for special missions requires that the exterior noise be minimized. Likewise, the interior noise for missions that transport combat troops or sensitive instrumentation must be controlled through accurate acoustic analysis. In accomplishing these tasks, Lockheed used noise prediction and control methods developed by Langley.

Aerodynamic Technologies

The aerodynamic development of the C-130 by Lockheed included the utilization of advanced databases of Langley experimental and computational studies. Although no wind-tunnel tests of the C-130 configuration have been conducted in Langley facilities, discussions of the interpretation and application of Langley aerodynamic concepts and data to the C-130 were common during the development of the aircraft. For example, Lockheed's interest in providing short takeoff and landing capabilities to the C-130 stimulated discussions on all facets of Langley research on high-lift systems, including trailing-edge flaps, propeller-induced effects, and powered-lift concepts. Lockheed remained particularly informed and interested in potential applications of the externally blown flap (EBF) and the upper-surface blowing (USB) powered-lift concepts conceived and developed at Langley. Numerous discussions with Langley researchers took place, especially with John P. Campbell and Joseph L. Johnson, Jr. for the EBF concept, and with Johnson for the USB concept.

Lockheed has applied computational codes developed by Langley to assess and improve aircraft characteristics, including the stall characteristics of the C-130J. Aerodynamic loads and several other significant aerodynamic topics have been investigated with the computational code USM3D, which was developed by Neal T. Frink of Langley. Another Langley aerodynamic code, FUN2DI, was used by Lockheed to simulate the effects of leading-edge ice shapes on the C-130J wing and empennage lifting characteristics. In this particular application, the code significantly reduced the number of tests required to support Federal Aviation Administration (FAA) certification.

Recognition Visit

As a gesture of goodwill and recognition of the contributions of all the NASA research centers (Langley, Ames, Glenn, and Dryden) to the development of the C-130 Hercules, Lockheed Martin arranged for a visit of a C-130J to the Langley Research Center on November 6, 1998. The C-130 flew to Langley from the Lockheed Martin facility in Marietta, Georgia. Langley Director Dr. Jeremiah Creedon welcomed representatives from Lockheed Martin, Congress, NASA Headquarters, and the Air Force Air Combat Command, as well as Langley employees in a formal ceremony at the Langley flight hangar. Mr. James A. (Micky) Blackwell, Jr., President of Lockheed Martin's Aeronautics Sector, recognized the contributions of NASA employees to the C-130 and the fruitful partnership between NASA and industry that it represented. Mr. Blackwell's compliments were especially meaningful to the Langley staff because he began his aerospace career as a NASA employee at Langley in 1962 and worked with Dr. Richard Whitcomb at the Langley 8-Foot Transonic Pressure Tunnel.

Langley researcher Mary Beth Wusk and her daughter complete their tour of the C-130J at Langley in 1998.

Lockheed Martin C-141 Starlifter

SPECIFICATIONS

Manufacturer
  Lockheed
Date in service
  May 1964
Type
  Transport
Crew
  Six
Engine
  Pratt & Whitney TF33-P-7

USERS

U.S. Air Force (Active, Air
National Guard, and Reserve)

DIMENSIONS

Wingspan . . . . 160.0 ft
Length . . . . 168.3 ft
Height . . . . 39.6 ft
Wing area . . . . 3,228 sq ft

WEIGHT

Empty . . . . 140,882 lb
Gross . . . . 343,000 lb

PERFORMANCE

Max speed . . . . . Mach number
                          of 0.83
Range . . . . 2,550 n mi

HIGHLIGHTS OF RESEARCH BY LANGLEY FOR THE C-141

1. Parametric tests in the Langley 16-Foot Transonic Dynamics Tunnel (TDT) of an isolated-tail model advanced the understanding of T-tail flutter at transonic conditions.

2. Although the tail design did not flutter in initial TDT tests, the parametric tests identified several issues, including a large area of transonic flow separation at the juncture of the vertical and horizontal tails that promoted early flutter for some configurations.

3. When an unpredicted elevator-induced flutter problem in flight was exhibited, TDT tests identified key factors and solutions.

4. The isolated-tail tests permitted flutter clearance for a complete model of the C-141 in subsequent TDT tests.

5. Langley developed a technique to test dynamically scaled models in TDT to assess an aileron reversal problem.

6. Joint Langley and Air Force runway performance tests with a C-141 aircraft established grooving as an effective solution to hydroplaning accidents.

The Lockheed (now Lockheed Martin) C-141was the first jet aircraft designed to meet military standards as a troop and cargo carrier. It was also the first military aircraft to be developed with a requirement for FAA type certification in the contract. The Starlifter is the workhorse of the Air Mobility Command. It fulfills a vast spectrum of airlift requirements through its ability to airlift combat forces over long distances, place those forces and their equipment either by conventional landings or airdrops, resupply employed forces, and extract the sick and wounded from a hostile area. The current C-141B is a stretched version of the original C-141A with in-flight refueling capability. The C-141B is about 23 ft longer than the C-141A, with cargo capacity increased by about one-third. The C-141 force, nearing seven million flying hours, has a proven reliability and long-range capability.

Several critical flutter tests for the C-141 were conducted in the Langley 16-Foot Transonic Dynamics Tunnel (TDT). During these tests, parametric experimental and computational research was conducted on the flutter characteristics of T-tail configurations with full-span models and unique isolated-tail models of the C-141 configuration. Studies of the aeroelastic mechanisms responsible for inducing flutter led to significant improvements in the prediction and analysis of T-tail flutter at transonic conditions and the development of specific modifications to cure flow separation near the intersection of the vertical- and horizontal-tail surfaces. During early service, the C-141 exhibited tail flutter that was precipitated by large elevator deflections. This phenomenon was also replicated and analyzed in the TDT. Langley also conducted research in the TDT to assess tendencies of the C-141 to exhibit the aeroelastic phenomenon known as high-speed "aileron reversal".

The C-141 aircraft was a test subject during joint NASA and Air Force runway traction studies, which demonstrated the effectiveness of a runway grooving concept developed by Langley. Fifty grooved and ungrooved runway surfaces were tested and grooving was established as an effective solution to hydroplaning accidents. Subsequently, runways and highways worldwide were grooved for better traction in inclement weather.

LANGLEY CONTRIBUTIONS TO THE C-141

Tail Flutter

In the 1950's, several domestic and foreign high-performance aircraft configurations were developed that incorporated T-tail empennage configurations. Unfortunately, operational experiences with some of these aircraft indicated that tail flutter could be a critical problem for this configuration at high subsonic speeds. In July 1954, a British Handley Page Victor bomber experienced severe T-tail flutter and crashed during a low-altitude, high-speed flight test when the tail began wobbling and then tore off the aircraft, which dove into the ground. Noted aviation periodicals, including *Aviation Week*, stated that the accident raised doubts on the high placed T-tail configuration (ref. 6). In addition, jet-powered T-tail flying boats had been developed for the Navy that also experienced flutter problems.

The staff of the Langley 16-Foot Transonic Dynamics Tunnel (TDT) was cognizant of the T-tail problem as the C-141 configuration entered the developmental process. The complex aeroelastic and aerodynamic factors that had allowed tail flutter to occur were discussed and researched in considerable depth by TDT staff members Frank T. Abbott, Charles L. Ruhlin, Maynard C. Sandford, and E. Carson Yates, Jr.. The Langley team and their peers recognized that the flutter mechanisms of T-tails were not well understood, particularly at critical transonic flight conditions where flutter margins were minimal. In addition, they identified certain geometric parameters, such as dihedral and sweep of the horizontal-tail surfaces, to be particularly critical. Carson Yates conducted extremely complex analytical calculations that identified the critical parameters and flutter modes for T-tails, and the results of his calculations highlighted the dramatic flutter challenge for transonic conditions.

A formal request from the Air Force for flutter clearance tests of the C-141 in the Langley TDT was received. In view of the relatively poor understanding of T-tail flutter, the Air Force, Lockheed, and Langley team agreed to conduct special tests of a dynamically scaled model of the empennage and rear fuselage in the TDT. These tests

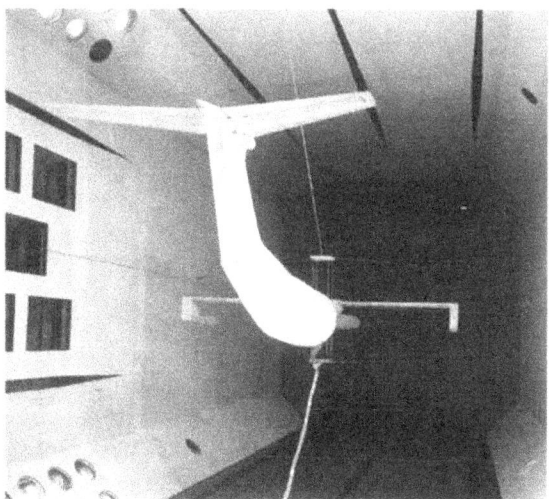

Flutter model of C-141 empennage in the Langley 16-Foot Transonic Dynamics Tunnel in 1961 with highly streamlined original bullet fairing at the juncture of the horizontal and vertical tails.

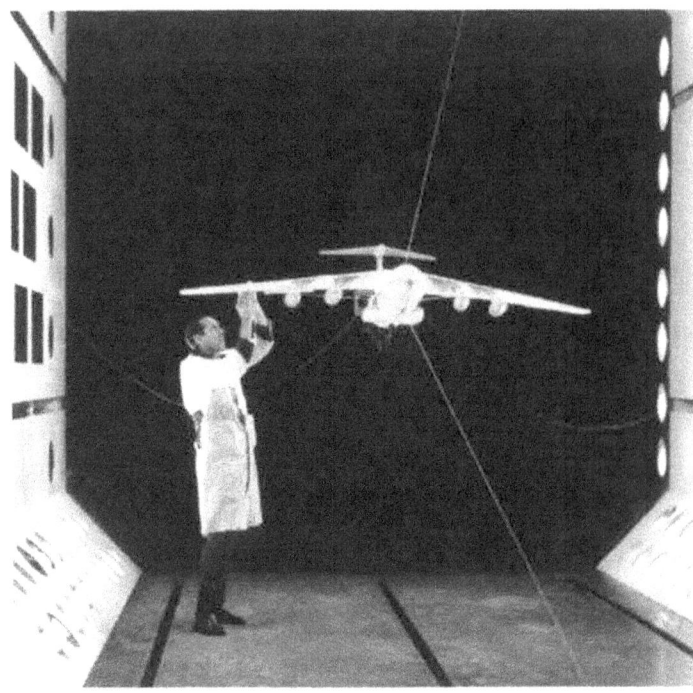

C-141 model being prepared for flutter tests in the Transonic Dynamics Tunnel in 1962.

permitted parametric studies and correlation with theory before the full-span model flutter clearance test. Flutter tests of this unique C-141 empennage model began in the TDT in late 1961. The design configuration did not flutter within the Mach number and dynamic pressure ranges tested. A matrix of structural parameters and test conditions was covered to determine the relative sensitivity of flutter to variations in physical features such as the stiffness of the horizontal-tail pitch trim actuator, stiffness of the fin spar, roll and yaw stiffness of fin-stabilizer joints, rotational stiffness of elevators and rudder, and stabilizer mass and yaw and roll.

During the isolated-tail tests, the NASA and industry team encountered an unexpected flow phenomenon on the tail at transonic speeds. The C-141 tail had been designed with a streamlined bullet-shaped fairing to improve airflow at the juncture of the vertical and horizontal tails. However, at transonic speeds the bullet shape caused adverse pressure gradients and shock-induced flow separation over the aft portion of the fin-stabilizer juncture. Tufts on the vertical tail illustrated the dynamic and massive flow separation regions as the tail fluttered over relatively large amplitudes. The flow separation acted as a forcing function to encourage certain flutter modes of the entire tail surface. Yates recommended a redesign of the bullet, including a nonstreamlined shape and a blunt "boat tail". Vortex generators on the fin were also evaluated. With the vortex generators and bullet modifications, results of the test indicated that the vertical-tail flow separation was entirely eliminated and any tendency for flutter was pushed significantly beyond the flight envelope.

The results of the isolated-tail model tests significantly increased the understanding of T-tail flutter prediction, and provided extremely valuable background for the upcoming flutter clearance test of the complete configuration. When the full-span flutter clearance model of the C-141 was tested in the TDT in 1962, no flutter was encountered within the flight envelope (flutter was pushed beyond 120 percent of design dive speed), and the value of the isolated-tail model tests was recognized by the participants.

After the C-141 entered service, an unexpected unstable oscillation of the horizontal tail was experienced during high-altitude flight tests. The flutter was precipitated by a deflection of about 8 deg of the elevator relative to the stabilizer; no fix was found in flight tests. Analysis by Lockheed indicated that the flutter speed could be raised significantly by increasing the elevator mass balance, and this approach was used to eliminate the problem. This elevator-induced flutter mechanism was studied by Maynard Sandford and Charles Ruhlin with the isolated-tail model in the TDT, where they found that the flutter was reproduced by the model. The tests also verified the effectiveness of elevator mass balancing to eliminate tail flutter.

In summary, Langley's contribution to the C-141 program in the area of flutter generated extensive advances in experimental techniques and analytical methods for fundamental understanding of T-tail flutter characteristics. The highly successful flutter studies of the C-141 resulted in significant advances in the development of T-tail configurations.

Aileron Reversal

Evaluations of the C-141 indicated that for certain flight conditions the aircraft experienced the phenomenon known as aileron reversal. This phenomenon occurs when a deflection for roll control of the aileron at the wing trailing edge results in aeroelastic twisting of the wing to the extent that the control effectiveness is nullified or actually reversed. That is, a pilot's input to intentionally roll the aircraft results in little or no response. In some cases, the effects of aileron reversal can actually roll the aircraft in the direction opposite to that intended by the pilot. Accurate predictions of this phenomenon require precise estimates of the aerodynamic behavior of the wing and aileron system under dynamic conditions and at high speeds. The difficulty of estimating aerodynamics at transonic conditions made this problem a significant challenge for the designers of emerging large, flexible transports such as the C-141 and C-5.

C-141 model in Transonic Dynamics Tunnel for aileron reversal tests in 1966.

Langley researcher Irving Abel developed an innovative experimental technique to evaluate aileron reversal tendencies with an aeroelastically scaled C-141 model in the TDT during 1964 and 1966. Abel used a cable-mount system in the TDT that had been developed by Wilmer H. Reed, Jr. and Frank Abbott to permit more realistic flutter tests by allowing the structural modes of the models to be simulated in free-flight conditions. Abel's results for the C-141 model agreed well with flight results and provided a new test method to the capabilities of the TDT.

Runway Grooving

In the 1950's, Langley researchers at the Langley Aircraft Landing Dynamics Facility focused their research on aircraft braking and directional control performance on wet runway surfaces. This effort was led by Langley researchers Walter Horne and Thomas J. Yager. The phenomenon of tire hydroplaning was identified as a contributory factor for unsatisfactory tire traction on wet runways. The associated losses in tire traction in water were defined for a variety of test parameters. A technique that appeared promising for improved water drainage at the tire and pavement interface was to modify the pavement surface with slots, or grooves, similar to the grooves in aircraft tire treads. Basic parametric research on the groove concept was conducted at the Aircraft Landing Dynamics Facility. The very positive results led to full-scale, instrumented aircraft tests at the landing research runway at the NASA Wallops Flight Facility in 1968.

In a joint program with the Air Force, Langley evaluated the effects of 50 grooved and ungrooved runway surfaces on the braking performance of a C-141 aircraft. On the basis of these test results and other aircraft evaluations, the application of grooving to runways and the nation's highways has been accepted as an efficient means for minimizing wet pavement skidding accidents.

C-141 braking tests in a joint NASA, Air Force, and FAA program at Wallops Flight Facility.

Wind Tunnel and Flight Correlation

During early flight tests of the C-141, the wing pressures and pitching moments were found to be considerably different from those predicted in wind-tunnel tests at supercritical Mach numbers. These discrepancies resulted in several national efforts to establish reasons for the differences, with emphasis on wind-tunnel test procedures. At Langley, Dr. Richard T. Whitcomb participated in the analysis of the data discrepancy and concluded that wind-tunnel scale effects were the problem. Under the direction of Whitcomb, Donald L. Loving led a study in 1965 that focused on the impact of

*C-141 model with dorsal strut and lower strut support in the
8-Foot Transonic Pressure Tunnel for drag correlation studies.*

shock-induced boundary-layer separation. Differences in the relative magnitudes of shock-induced flow separation resulted in large differences in aerodynamic loads on the wing. Unfortunately, the phenomenon was difficult to model and extremely complex. The problem was differences between the relative thickness of boundary layers on models and full-scale aircraft. The importance of artificially locating transition on a model to produce the same relative boundary-layer thickness at the trailing edge of the wing was highlighted. However, locating the transition point at the appropriate place on the model proved to be more an art than a science. The major recommendation from the study was to conduct wind-tunnel tests for several artificial positions of transition to evaluate the sensitivity of shock-induced separation to modifications of the boundary-layer conditions. Later, James A. (Micky) Blackwell, Jr. of Langley contributed an approach to scaling and transition-fixing that was accepted as producing accurate simulations for wind-tunnel practice.

In a tunnel-to-flight study, Langley awarded a contract to Lockheed in 1970 to conduct analytical studies on drag estimation for the C-141 and to obtain a set of fully corrected wind-tunnel data on a 0.0275-scale C-141 model in the Langley 8-Foot Transonic Pressure Tunnel. Lockheed's experience in the C-5 program with new transition-fixing techniques and model support systems was applied to the C-141 test data to obtain the accuracy required for the study. Model support interference corrections were evaluated through a systematic series of tests, and the fully corrected model data were analyzed to provide details of the model component interference factors. The results of the investigation indicated that the predicted, subcritical minimum profile drag of the complete C-141 configuration was within 0.7 percent of flight-test data.

Lockheed Martin F-16 Fighting Falcon

SPECIFICATIONS

Manufacturer
  Lockheed Martin
Date in service
  January 1979
Type
  Fighter-attack
Crew
  One or Two
Engine
  F-16C . . . Pratt & Whitney
  F100-PW-200
          . . . General Electric
  F110-GE-129

USERS

U.S. Air Force, U.S. Navy,
Bahrain, Belgium, Denmark,
Egypt, Greece, Indonesia, Israel,
South Korea, Netherlands,
Norway, Pakistan, Portugal,
Singapore, Taiwan, Thailand,
Turkey, and Venezuela

DIMENSIONS

Wingspan. . . . . 31.0 ft
Length . . . . 49.3 ft
Height . . . . 16.7 ft
Wing area . . . . 300.0 sq ft

WEIGHT

Empty . . . . .18,591.0 lb
Typical combat . . . .23,498.0 lb

PERFORMANCE

Max speed . . . . . . .above Mach
                number of 2.0
Radius of action
  Air patrol . . . . 866 n mi
  Typical strike . . . . 676 n mi

HIGHLIGHTS OF RESEARCH BY LANGLEY FOR THE F-16

1. At the request of industry and Department of Defense, Langley staff participated in F-16 reviews, from the YF-16 prototype to the current F-16 variants.

2. The F-16 uses Langley wind-tunnel research on leading- and trailing-edge flap technology for transonic maneuver optimization.

3. Langley's fly-by-wire technology and sidestick controller are incorporated in the F-16.

4. Lockheed Martin Tactical Aircraft Systems (LMTAS) sought Langley's guidance and accepted recommendations for a wing-body strake that provides superior maneuverability through the aerodynamic phenomenon known as vortex lift.

5. Langley helped solve numerous F-16 developmental challenges including

    *   Flutter clearance

    *   Spin recovery

    *   High-angle-of-attack stability and control

    *   Recovery from a deep stall

6. Langley provided supersonic wing design methods and developmental tests in a cooperative program with LMTAS for the F-16XL supersonic cruise prototypes.

The conception, development, and deployment of the F-16 by the Lockheed Martin Tactical Aircraft Systems (LMTAS) Division (formerly General Dynamics) included a close working relationship with Langley Research Center. Throughout the design and deployment of the F-16, Langley personnel participated in evaluations, wind-tunnel tests, and special studies—all of which helped to insure superior maneuverability and performance for this first-line U.S. fighter-attack aircraft. The LMTAS and NASA partnership began with the YF-16 Lightweight Fighter (LWF) Program in 1971. The YF-16 concept of a relatively small, highly maneuverable day fighter ultimately evolved into the multimission, all-weather F-16, which has been deployed worldwide by numerous countries. General Dynamics was awarded the Collier Trophy for 1975 in recognition of accomplishments in the development of the F-16. The F-16, with its primary mission as a strike aircraft, complements the Air Force F-15 air superiority fighter. The aircraft has been a workhorse for several nations in recent conflicts including Operation Desert Storm, Bosnia, and Kosovo.

Designers of the F-16 frequently interacted with the Langley staff and used Langley facilities during the development cycles for the YF-16 and the F-16. Existing NASA data and additional cooperative tests were used to optimize aerodynamic performance, stability and control, and aeroelasticity. Langley facilities used by the LMTAS and NASA team included the 30- by 60-Foot (Full-Scale) Tunnel, the 20-Foot Vertical Spin Tunnel, the Unitary Plan Wind Tunnel, the 16-Foot Transonic Dynamics Tunnel, the 7- by 10-Foot High-Speed Tunnel, and the Differential Maneuvering Simulator.

Langley was involved in research on several variants of the F-16, including the highly impressive F-16XL, which was a derivative that can cruise efficiently at supersonic speeds without use of an afterburner. Although not put into production, the F-16XL demonstrated the validity of Langley design methodologies for supersonic cruise vehicles.

LANGLEY CONTRIBUTIONS TO THE F-16

Early Assessments

In early 1972, the Langley Research Center was requested by the Department of Defense (DOD) to participate in assessments and tests of competing YF-16 and YF-17 designs for the Air Force Lightweight Fighter (LWF) Program. Langley researchers, as members of DOD source evaluation teams, assessed technical claims of each of the competing contractors. The YF-16 configuration underwent extensive wind-tunnel tests at Langley, especially in the 30- by 60-Foot (Full-Scale) Tunnel, the 20-Foot Vertical Spin Tunnel, the 7- by 10-Foot High-Speed Tunnel, the 16-Foot Transonic Dynamics Tunnel, and the Unitary Plan Wind Tunnel. Extensive studies of enhanced control systems for high-angle-of-attack conditions were also conducted in the Langley Differential Maneuvering Simulator (DMS) with pilots from Langley, Lockheed Martin Tactical Aircraft Systems (LMTAS) Division, and the Air Force. The results of these studies provided information on the capabilities of the YF-16. Following the selection of the YF-16 for production as the F-16 in 1974, the precursor tests served as an excellent knowledge base for development of the new F-16 fighter.

Transonic Performance

The air war experience in Vietnam, where the lack of maneuverability of U.S. fighters at transonic speeds provided advantages to nimble enemy fighters, was the stimulus for the YF-16 program. The Air Force and designers of the YF-16 therefore placed great emphasis on achieving unprecedented transonic maneuver capability with excellent handling qualities. At that time, Langley researchers under the leadership of Edward C. Polhamus were conducting in-depth studies in the 7- by 10-Foot High-Speed Tunnel of approaches to obtain near optimum aerodynamic maneuver performance for wings, including the use of fixed and variable-camber concepts. Some of the earliest systematic wind-tunnel tests determined the most effective geometries for leading- and trailing-edge wing flaps. In addition, studies were conducted to develop methodologies for the prediction and minimization of undesirable buffet characteristics. The program was coordinated with flight tests of actual high-performance fighters at the NASA Dryden Flight Research Center.

Polhamus and his group (including Edward J. Ray, Linwood W. McKinney, Blair B. Gloss, and William P. Henderson) provided valuable guidance to the LMTAS design team. The insight and understanding provided by the broad database from Langley tests permitted development of the extremely effective leading- and trailing-edge flaps used by the F-16. The F-16 (and most other high-performance fighters) uses specific schedules of flap deflection with Mach number and angle of attack for superior maneuverability at transonic conditions.

Vortex Lift

In the early 1960's worldwide interest in the phenomenon known as "vortex lift" increased as a result of aerodynamic studies of highly swept configurations such as the Concorde supersonic transport. Two events contributed to the initiation of a world-class Langley vortex-lift research program led by Edward Polhamus.

The first event was a detailed experimental and theoretical study of canard configurations at high subsonic speeds led by researcher Linwood McKinney. McKinney became interested in the favorable effects of vortex on lift that were demonstrated during development of the Swedish Viggen canard configured aircraft. McKinney's study indicated

F-16 in a hard turn with wing leading-edge flaps deflected and vortices produced by the sharp wing-body strakes illustrated by condensation.

that the favorable effects of the canard trailing vortex on the lifting capability of a close-coupled wing might also be extended to higher angles of attack by the strong leading-edge vortex flow of a slender lifting surface.

The second event that led to the vortex-lift work at Langley was a cooperative Langley and Northrop study of hybrid wings that centered on the use of relatively large, highly swept wing extensions at the wing-fuselage intersection to promote strong beneficial vortex-flow effects.

Stimulated by the promise of this revolutionary aerodynamic concept, Polhamus and his associates put together a vortex-lift research program that became internationally recognized for its experimental database, analytical procedures, and aircraft applications. In addition to Polhamus and McKinney, key members of this team included Edward Ray, William Henderson, John E. Lamar, and James M. Luckring. Their research into the fundamentals and applications of vortex flows allowed Langley to aid U.S. industry in the design of highly maneuverable advanced fighters.

The evolution of the YF-16 design at LMTAS included studies of configuration variables such as wing design, maneuvering devices, number and location of engines, control surfaces, number and location of tail surfaces, and structural concepts. As the configuration options matured, two candidate configurations competed for priority. The first configuration was a simple wing, body, and empennage design, while the second design was a twin-tailed, blended-wing body with vertical and horizontal tails on booms. The LMTAS team selected the best features of both configurations for the final YF-16 design. After considerations of performance, stability, and control were addressed, the YF-16 configuration incorporated a rather wide, blended forebody that produced strong vortices at moderate angles of attack. LMTAS had attempted to weaken the strength of the vortices by promoting attached flow, but these attempts were not successful.

A team from LMTAS visited the Polhamus group and requested guidance for control of the vortical flows emanating from the forebody of the YF-16. In a historically significant meeting, the Langley team suggested that a completely different approach be used to control the vortical flow. Specifically, Langley suggested that the leading edge of the blended forebody be sharpened to increase (rather than decrease) the strength of the vortices, which could be exploited for additional lift. This modification allowed the forebody vortices to dominate and stabilize the flow field over the aircraft at high angles of attack, improve longitudinal and directional stability for the single-tail configuration, and stabilize the flow over the outer wing panels. The LMTAS team accepted the recommendation, and subsequent wind-tunnel tests verified the lift-enhancing effect of the sharpened wing-body strake. In addition, the sharpened strake significantly reduced buffet intensity at transonic maneuvering conditions. The wing-body strake of the F-16 is regarded as a key contribution to its success as a maneuvering fighter.

When the YF-16 team analyzed the effects of deflected leading- and trailing-edge flaps and the sharp-edged wing-body strake on directional stability at high angles of attack, they found that the stability contributions of a single vertical tail were significantly enhanced. However, the contributions of twin vertical tails were markedly degraded. As a result of this analysis, the YF-16 was configured with a single vertical tail. Thus, the Langley recommendation for a sharpened wing-body strake favorably impacted other configuration features of the aircraft.

High-Angle-Of-Attack Stability and Control

Increased maneuverability for the YF-16 necessitated extended flight at high angles of attack where aerodynamic deficiencies caused by separated airflow can result in sudden decreases in stability and controllability. Therefore, special emphasis was placed on tests in Langley facilities to insure that the YF-16 could provide the pilot with "care-free" maneuverability. Langley facilities used included the Full-Scale Tunnel, the Spin Tunnel, and the DMS. Helicopter drop-model tests for the YF-16 and the F-16 were not conducted at Langley because of concurrent resource demands of the B-1 and F-15 programs and the fast pace of the LWF Program.

Researcher William A. Newsom, Jr. with the free-flight model of the YF-16 used in tests in the Full-Scale Tunnel.

To provide superior handling characteristics at high angles of attack, any undesirable handling characteristics were pushed out of the operating envelope of the aircraft and the flight envelope was limited with an advanced fly-by-wire flight control system by LMTAS. This concept has proven to be highly successful and has been used in all variants of the F-16.

Researchers in the Full-Scale Tunnel conducted exhaustive tests of the YF-16 and the F-16 configurations at high-angle-of-attack conditions. Data gathered in these studies identified the aerodynamic contributions of components of the airframe, deficiencies in stability and control, and potential solutions to these deficiencies. These data also formed the basis for piloted simulator studies at Langley and LMTAS that helped LMTAS design the flight control system for critical high-angle-of-attack conditions.

Langley and LMTAS engineers recognized that reliance on the flight control system to insure satisfactory behavior at high angles of attack required research on the ability of fly-by-wire control systems to limit certain flight parameters during strenuous air combat maneuvers. The YF-16 and F-16 employ the concept of "relaxed static stability" in which the aircraft is intentionally designed to be aerodynamically unstable while the flight control system provides integrated stability by sensing critical flight variables and making the control inputs required to stabilize the aircraft. Cooperative piloted simulator studies were conducted in the Langley DMS to identify critical control system components, schedules, and feedback gains to stabilize the aircraft and pilot system for the most demanding maneuvers for high-angle-of-attack conditions. Of particular concern was the ability of the horizontal tails and longitudinal control system to limit the aircraft's angle of attack during maneuvers with high roll rates at low airspeeds. Such maneuvers are critical because rapid rolling maneuvers produce large nose-up trim changes due to inertial effects, whereas the aerodynamic effectiveness of the horizontal tails becomes significantly reduced at low airspeeds and high angles of attack.

Under the leadership of Langley researchers Luat T. Nguyen and William P. Gilbert, extensive studies were conducted in the DMS with pilots from Langley, LMTAS, and the Air Force. These studies verified the effectiveness of the flight control system and identified critical maneuvers that would be tested during the flight development program. Gains in the flight control system were modified and incorporated into the aircraft system. New control elements, such as a yaw rate limiter, a rudder command fade-out, and a roll rate limiter that maximized the maneuvering envelope with minimal adverse effects were developed. The DMS was flown by the LMTAS YF-16 and F-16 test pilots who later flew the prototypes. These pilots cited the accuracy of the preflight predictions and the valuable training and exposure to potentially hazardous flight-test conditions provided by the DMS.

Curing Deep Stall

Early in Langley research, tests of a YF-16 model in the Full-Scale Tunnel indicated that if angle of attack was not limited by the flight control system, the aircraft could pitch up and attain an undesirable trimmed condition at very high angles of attack with insufficient nose-down aerodynamic control to recover normal flight. The Langley researchers viewed this "deep" stall as a serious problem that would require significant research for resolution. The ability of the YF-16 to enter this deep stall was demonstrated in piloted simulation with the Langley DMS and the results were formally reported to the LMTAS team and the Air Force. Unfortunately, other NASA and industry wind-tunnel tests of YF-16 models contradicted the Full-Scale Tunnel results—indicating that the problem would not exist. The engineering community generally agreed that the deep stall was not an issue for the F-16. Flight tests of the YF-16 aircraft were not extensive enough to determine susceptibility to deep-stall phenomenon. However, the follow-on flight-test program for the F-16 proved that the favorable projections were wrong and that the deep-stall condition actually existed for the aircraft.

High-angle-of-attack test results obtained on the early production version of the F-16 configuration in the Full-Scale Tunnel showed the same deep-stall trimmed condition that was noted in the YF-16 results. Again, contradictory wind-tunnel results obtained elsewhere convinced the engineering community that the aircraft would not exhibit the problem. However, in subsequent high-angle-of-attack flight evaluations at Edwards Air Force Base, an F-16 that had been subjected to rapid rolls at diminishing airspeeds in vertical zoom climbs suddenly entered a stabilized deep-stall condition and the pilot was unable to recover the aircraft with normal aerodynamic controls. The conditions for the deep stall agreed very well with predictions based on the earlier test results from the Langley Full-Scale Tunnel. Fortunately, the test aircraft was equipped with an emergency spin recovery parachute that was deployed to recover the aircraft to normal flight conditions. This event brought all high-angle-of-attack flight tests of the F-16 to a standstill while a solution to the deep stall could be found.

The excellent correlation of the Langley wind-tunnel data with the events that occurred in flight and the ongoing evaluation of the F-16 in the DMS at Langley provided the tools to work the problem. A joint NASA, LMTAS, and Air Force team arrived at Langley and aggressively sought interim and permanent fixes under Nguyen's lead. Working with the evaluation pilots, the team devised a "pitch rocker" technique in which the pilot pumped the control stick fore and aft thereby setting up an oscillatory pitching motion that forced the aircraft out of the deep stall and allowed recovery to normal flight. The concept was adopted and validated in the F-16 flight-test evaluation at Edwards and was incorporated in the flight control system as a pilot selectable emergency mode. A longer term fix was developed in cooperative wind-tunnel tests in the Full-Scale Tunnel and in LMTAS wind tunnels. The ultimate fix for the problem (which also improved takeoff performance) was increasing the size of the horizontal tail about 25 percent. This solution has been incorporated in all F-16 production aircraft.

Langley also identified and developed an automatic spin prevention control system concept that could prevent inadvertent spins for the F-16. Nguyen and his group demonstrated the effectiveness of the system with the DMS and analytical studies. The fundamental elements of the system were further refined by LMTAS and incorporated in the F-16 fleet.

As a result of Langley supporting analysis and tests, the advanced fly-by-wire flight control systems of the YF-16 and F-16 were designed with confidence that inadvertent loss of control incidents that had plagued many earlier fighters would be eliminated.

Spin Recovery

The YF-16 and F-16 configurations underwent extensive tests in the Langley Spin Tunnel to determine spin and spin recovery characteristics, especially for the configurations with external store loadings. In addition, these tests determined the size of emergency spin recovery parachute that is required for the flight trials. The results of these tests indicated trends that were fairly typical of high-performance fighters. The models predicted two types of spins—one spin was a relatively well-behaved moderate spin from which recovery was easily effected, and the second spin was a potentially serious flat spin at very high angles of attack with poor recovery. The success of the sizing of the spin recovery parachute was proven in its successful deployment in the unexpected deep-stall encounter. The parachute can be credited with saving the invaluable test aircraft as well as the pilot. The development and implementation of the production automatic spin prevention system by the Langley and LMTAS team has minimized operational encounters with spins.

Fly By Wire and Sidestick Controller

In 1954, flight tests of the first fly-by-wire aircraft, a modified F9F Panther jet, were initiated at Langley. The primary objective of the tests was to evaluate various automatic control systems, including those based on rate- and normal-acceleration feedback. However (as is the case in many research investigations), the most valuable result of the flight test was related to secondary objectives—in this case the introduction and evaluation of fly by wire and a sidestick controller for pilot inputs.

In a serendipitous approach, Langley researchers decided to avoid the relatively large expense and time required to modify the existing hydraulic flight control system for the F9F. Instead they chose to implement an auxiliary system based on a fly-by-wire analog concept and a small (4 in.) sidestick controller mounted at the end of the armrest at the side of the pilot. The sidestick controller was used as the maneuvering flight controller throughout the investigation. Rapid and precision maneuvers such as air-to-air tracking, ground strafing runs, and precision landings were evaluated.

Langley test pilot William L. Alford checks out the side-stick controller in 1954 studies.

The objectives of the flight program were completed with great success, and the information on various types of automatic control feedback was used for numerous aircraft development programs. However, the very successful use of the rudimentary fly-by-wire and sidestick controller concepts generated considerable excitement within the research community, especially those visionaries that anticipated the weight saving advantages for future aircraft. Additional research was conducted at Langley on these systems, including the use of a sidestick controller for the Apollo mission. The concept of the fly-by-wire control system was later refined and greatly improved in cooperative efforts between Langley and Dryden that used a modified Vought F-8 Crusader and surplus Apollo digital computers.

Flutter Clearance

Flutter clearance tests of both the YF-16 and F-16 were conducted in the Langley 16-Foot Transonic Dynamics Tunnel (12 tunnel entries for the F-16) under the leadership of Moses G. Farmer, Raymond G. Kvaternik, Jerome T. Foughner, and Frank W. Cazier. Numerous external store configurations were investigated and several potential flutter problems were identified and solved. LMTAS requested a large number of flutter tests to identify critical conditions that would be demonstrated in flight, thereby minimizing very expensive flight-test time. The YF-16 and F-16 did not demonstrate flutter within their flight envelopes in the tunnel tests, but valuable improvements in flutter margins were identified. For example, a fuel-usage sequence was established for the three-bay external tanks that significantly raised the flutter speed. Mass balancing of the wingtip missile launcher corrected a potential flutter problem with the wingtip missile and underwing missile combination. Several TDT studies were directed at active flutter suppression for the F-16 with stores. Beginning in 1979, a joint NASA, General Dynamics, and U.S. Air Force Wright Aeronautical Laboratories team initiated a series of tests that continued over an 8-year period. These highly successful studies progressed through an evaluation of a digital adaptive (no prior knowledge of the aircraft configuration) system. A Langley concept for flutter suppression known as the "decoupler pylon" was designed and evaluated by Wilmer H. Reed, Jr., Cazier, and Farmer, including flight tests on an F-16 aircraft; however, the concept was not implemented in the F-16 fleet.

Jerome T. Foughner checks out the F-16 model with external stores mounted in the 16-Foot Transonic Dynamics Tunnel for flutter tests.

The F-16XL.

In the mid-1970's the U.S. Air Force became interested in a fighter aircraft capable of "supercruise"—the ability to cruise supersonically without an afterburner while retaining respectable maneuver, takeoff, and landing characteristics. The supercruise requirement drove aircraft configurations to highly swept wing platforms. LMTAS appreciated the fact that the modular construction of the YF-16 allowed for relatively simple replacement of the outer wing panels and that a supercruiser demonstrator aircraft with a highly swept wing would undoubtedly attract considerable interest within the Air Force.

Langley's staff had developed a research program known as the Supersonic Cruise Integrated Fighter (SCIF) Program under the leadership of Roy V. Harris, Jr. As participants in previous national and NASA civil supersonic transport programs (SST), the Langley staff were leaders in the development of databases and design methods for efficient SST configurations. Several in-house supercruiser fighters were designed and tested across the speed ranges at Langley. Subsequent to the SCIF program, Langley joined several industry partners in cooperative, nonproprietary studies of supercruiser configurations.

In 1977 Langley and LMTAS agreed to a cooperative study to design a new cranked-arrow wing for the F-16 to permit supersonic cruise capability. Personnel from LMTAS worked alongside the NASA researchers under the direction of Charles M. Jackson at Langley during the studies. The project leader for supersonic design was David S. Miller. The results of the wind-tunnel and analytical studies indicated that a viable wing could be designed to satisfy the supersonic and transonic requirements. With these results, LMTAS initiated a company funded development of an F-16 derivative with supersonic cruise capability. Following the spirit of the previous wing design cooperative venture with NASA, a cooperative agreement was signed for mutual efforts on the new demonstrator, which was called the Supersonic Cruise and Maneuver Prototype (SCAMP).

Project engineer Sue B. Grafton with the free-flight model of the F-16XL.

F-16XL prototype in flight.

Extensive tests for SCAMP took place in Langley facilities, including the Unitary Plan Wind Tunnel, the 7- by 10-Foot High-Speed Tunnel, the 16-Foot Transonic Dynamics Tunnel, the Full-Scale Tunnel, the DMS, the Spin Tunnel, and a helicopter drop model. During these tests, a team led by researcher Joseph L. Johnson, Jr. identified low-speed stability and control issues that required modifying the wing apex with a rounded planform. Research on the SCAMP configuration by Langley researchers identified numerous advanced concepts for improved performance, including the application of vortex flaps on the highly swept leading edge for improved low-speed and transonic performance, automatic spin prevention concepts, and optimized wings for supersonic cruise. The final configuration became known as the F-16XL (later designated the F-16E), which displayed an excellent combination of reduced supersonic wave drag, utilization of vortex lift for transonic and low-speed maneuvers, low structural weight, and good transonic performance. The F-16XL flutter envelope was cleared in the 16-Foot Transonic Dynamics Tunnel by Charles L. Ruhlin without significant problems.

Two (a one-seat and a two-seat) F-16XL demonstrator aircraft were subsequently built and entered flight tests in mid-1982. In recognition of Langley's many contributions to the F-16XL, LMTAS management sent letters of recognition to Langley and senior NASA management. Marilyn E. Ogburn of Johnson's group was an invited participant at flight-test evaluations of the F-16XL at Edwards Air Force Base. The results of flight tests validated the accuracy of Langley wing design procedures, wind-tunnel predictions, and control system designs based on DMS tests. Unfortunately, the interest in supersonic cruise was replaced by an urgency to develop a dual role fighter with ground strike capability. Although the relatively large wing of the F-16XL carried a significant amount of weapons, the Air Force ultimately selected the F-15E in 1983 for developmental funding and terminated interest in the F-16XL.

Lockheed Martin F-22 Raptor

SPECIFICATIONS

Manufacturer
  Lockheed Martin
Type
  Air superiority fighter
Crew
  One
Engine
  Pratt & Whitney F119-PW-100

USER

U.S. Air Force

DIMENSIONS

Wingspan. . . . . 44.5 ft
Length . . . . 62.1 ft
Height . . . . 16.4 ft
Wing area . . . . .840 sq ft

WEIGHT

Empty . . . . 31,670 lb
Max takeoff . . . . 60,00 lb

PERFORMANCE

Max speed . . . . . . .above Mach
                number of 1.7

HIGHLIGHTS OF RESEARCH BY LANGLEY FOR THE F-22

1. At the request of industry and the Department of Defense, Langley personnel partic-ipated in reviews from the genesis of the Advanced Tactical Fighter (ATF) Program and the YF-22 Prototype Program to the production F-22.

2. The F-22 utilizes the results of Langley research and development of thrust-vector-ing nonaxisymmetric nozzles and afterbody integration for enhanced maneuverabil-ity and reduced drag.

3. Langley research of the high-angle-of-attack characteristics of the YF-22 and F-22 proved to be highly accurate and contributed to the outstanding agility and spin resistance of the configuration.

4. Supersonic drag and performance wind-tunnel measurements by Langley agreed within 1 percent with flight measurements, which ensured supersonic cruise without afterburner.

In the late 1980's, concern arose among military planners about the aging design (first flight in 1972) of the F-15 and the possible loss of future air superiority of the fighter. Soviet fighters such as the MiG 29 and Su-27 had demonstrated remarkable maneuverability and performance. In addition, fighter technology had taken enormous strides forward with the introduction of stealth, or low observable, technology. There was also growing concern over the increased effectiveness of the Soviet air defense system that posed a highly lethal environment for the F-15. Therefore, the Air Force initiated an Advanced Tactical Fighter (ATF) Program to develop a replacement for the F-15. The request for proposals (RFP) was sent to industry in September 1985, and Lockheed and Northrop were chosen in October 1986 to develop two prototype air vehicles designated YF-22 and YF-23, respectively. Lockheed teamed with Boeing and General Dynamics, while Northrop teamed with McDonnell Douglas.

The Air Force asked that Langley provide support for the YF-22 and YF-23 development programs on an as requested, equal basis. Within their program funds and interests, Lockheed requested Langley's support in two key areas: supersonic cruise performance and high-angle-of-attack and spin technology. Langley facilities used in the development of the YF-22 and the F-22 included the Unitary Plan Wind Tunnel, the 30- by 60-Foot (Full-Scale) Tunnel, and the 20-Foot Vertical Spin Tunnel. The ATF Program was initially conducted at the highest levels of classification, and Langley demonstrated once again that it could make significant contributions in a highly controlled environment.

Langley's contributions to the development and demonstration of the YF-22 were cited in a letter of appreciation from the Lockheed Vice President for ATF, James A. (Micky) Blackwell, Jr. to the Langley Center Director, Richard H. Petersen in March 1991. Blackwell praised the Langley efforts and support of the YF-22 and cited the accuracies of Langley wind-tunnel predictions and the dramatic demonstrations of the performance and agility of the prototype. Follow-on Langley support for the production F-22 was conducted in these same areas.

LANGLEY CONTRIBUTIONS TO THE F-22

The Advanced Tactical Fighter (ATF) Program

By the late 1980's, concern arose among military planners about the aging design (first flight in 1972) of the F-15 and the possible loss of future air superiority of the fighter. Soviet fighters such as the MiG 29 and the Su-27 had demonstrated remarkable maneuverability and performance, and were provided to Third World countries. Also, fighter technology had taken enormous strides with the introduction of stealth, or low observable, technology. There was growing concern over the effectiveness of the Soviet air defense system, because new surface-to-air missiles posed a highly lethal environment for the F-15. Therefore, the Air Force initiated an Advanced Tactical Fighter (ATF) Program to develop a replacement for the F-15. Throughout this planning and competitive period, members of the Langley staff were involved in briefings and discussions with Department of Defense (DOD), Air Force, and industry teams about technical issues and concepts to be exploited by the new aircraft. In October 1985, the Air Force issued the original ATF request for proposals (RFP) to seven competing companies. The next month, the Air Force revised the RFP to include more stringent low observable requirements. In June 1986, Lockheed, Boeing, and General Dynamics signed a teaming memorandum of agreement, but continued to compete individually for the ATF contract. Prime contracts were subsequently awarded by the Air Force to Lockheed and Northrop for the development of flying prototypes. Lockheed's design was designated the YF-22, and Northrop's design was the YF-23.

Value of the Langley Database

The YF-22 and F-22 aircraft benefited from basic and applied aeronautics research conducted at Langley over the 20-year period prior to the evolution of the programs. In particular, Langley research in nonaxisymmetric thrust-vectoring nozzles, supersonic cruise without afterburner, and "carefree" maneuverability with unlimited angle of attack had produced concepts, data, and assessments of advanced technologies that ultimately provided the YF-22 and F-22 with superior characteristics.

The staff of the Langley 16-Foot Transonic Tunnel had long been recognized as the world leaders in conceptual research on nonaxisymmetric thrust-vectoring nozzles and aft-end aerodynamic integration for advanced fighters. Fundamental studies of isolated nozzle performance and vectoring efficiency were coupled with specific configuration evaluations during the 1960's, 1970's, and 1980's to the point that risk was considerably reduced, along with dramatic predictions for improvements in fighter maneuverability, cruise performance, and takeoff and landing performance.

The contributions of Langley researchers in the prediction and reduction of supersonic drag were also well recognized, including fundamental theories and design methods to reduce supersonic wave drag. Langley researchers also participated in in-house and cooperative studies with industry on supersonic civil and military aircraft. These researchers, who used the supersonic capabilities of the Langley 4- by 4-Foot Supersonic Pressure Tunnel and the Langley Unitary Plan Wind Tunnel, conducted years of research that came to fruition in the F-22 Program.

The Langley research program on high-angle-of-attack characteristics of fighter aircraft, which was initiated in 1969 in response to an alarming number of fighter aircraft accidents due to deficient characteristics, provided proven test techniques, design concepts, and advanced technology for use by the YF-22 and F-22 design teams. The staffs of the Langley 30- by 60-Foot (Full-Scale) Tunnel and the Langley 20-Foot Vertical Spin

Tunnel worked closely with industry in generic and specific applications of their technologies.

The YF-22 Program

During the ATF competition, the Air Force established that each contracting team could draw upon NASA's unique expertise and facilities on an equal basis. The Lockheed YF-22 team intended to place considerable emphasis on achieving unlimited angle-of-attack capability and supersonic cruise for their design. Lockheed requested tests in the Langley Unitary Plan Wind Tunnel, Spin Tunnel, and Full-Scale Tunnel in support of this goal.

In mid-1989, static and dynamic tests of the YF-22 began in the Full-Scale Tunnel under the direction of Sue B. Grafton to determine stability and control at high angles of attack. Data was also obtained to develop aircraft control laws. Free-flight tests of a YF-22 model to determine low-speed longitudinal and lateral-directional response characteristics and departure resistance were also conducted. The Langley Spin Tunnel was used to determine the spin and recovery characteristics of the YF-22 in 1988 and to validate the geometry of the emergency spin recovery parachute. Tests were also conducted in the Spin Tunnel to determine rotational aerodynamics during simulated spinning conditions for use in the development of control laws.

The first flight of the YF-22 prototype occurred on September 29, 1990, and the flight-test phase of the competition ended on December 28, 1990.

Free-flight tests of the YF-22 model in the Langley Full-Scale Tunnel in 1990.

Letter of Appreciation

The success and quality of Langley's contributions to the YF-22 Program were empha-sized in a March 12, 1991, letter from James A. Blackwell (Vice President and General Manager of the Lockheed ATF Office) to Richard H. Petersen, Director of the Langley Research Center. In the letter, Blackwell stated

> *On behalf of the Lockheed YF-22 Team, I would like to express our apprecia-tion of the contribution that the people of NASA Langley made to our success-ful YF-22 flight test program, and provide some feedback on how well the flight test measurements agreed with the predictions from your wind-tunnel measurements.*
>
> *Briefly, the flight test program was an enormous success, and we achieved all of our objectives. Between first flight on September 29, 1990, and the last flight on December 28, 1990, we flew the two prototypes from 80 knots and 60° angle of attack to Mach 2+.*
>
> *We relied on the NASA Langley Unitary Tunnel to provide the final supersonic drag and stability and control data that were used for performance predic-tions and the flight control law design. We demonstrated supercruise on both the GE and P&W-powered prototypes as predicted. The supersonic drag determined from flight test agreed within one percent of the predictions that were made using the Langley Unitary Tunnel data. The supersonic flying qualities were rated excellent; we flew to 15° angle of attack supersonically; and we did 360° full-stick rolls demonstrating excellent roll rates and roll response.*
>
> *The highlight of the flight test program was the high-angle-of-attack flying qualities. We relied on: aerodynamic data obtained in the 30- by 60-Foot Full-Scale Wind Tunnel to define the low-speed, high-angle-of-attack static and dynamic aerodynamic derivatives; rotary derivatives from your Spin Tunnel; and free-flight demonstrations in the Full-Scale Wind Tunnel. We expanded the flight envelope from 20° to 60° angle-of-attack, demonstrating pitch atti-tude changes and full-stick rolls around the velocity vector in 7 calendar days, December 11 to December 17. The reason for this rapid envelope expansion was the quality of the aerodynamic data used in the control law design and pre-flight simulations.*
>
> *Without the help of the NASA Langley people and your unique test capabili-ties, it would have been impossible to complete our YF-22 flight test program in the time available. We look forward to working with you and your people when we win the Full-Scale Development contract to build the F-22.*

The Lockheed, Boeing, and General Dynamics team was announced as the winner of the ATF competition on April 23, 1991.

The F-22 Program

Although not immediately apparent to the uninitiated observer, the external geometry of the F-22 changed significantly from the YF-22 prototype. Specifically, the wingspan was increased, the wing leading-edge sweep was decreased, the vertical tails were reduced in area and moved aft, and the horizontal-tail surfaces were reconfigured. These changes were considered significant enough to warrant specific tests of the F-22 in the same unique Langley facilities as the YF-22.

High-angle-of-attack tests of the F-22 were conducted in the Full-Scale Tunnel in 1992, and spin and rotary-balance tests were conducted in 1993 in the Spin Tunnel. In the F-22 Program, a free-flight model was not built, and tests in the Full-Scale Tunnel were limited to conventional static force and dynamic (forced-oscillation) tests.

Sue Grafton with the F-22 force test model for the Langley Full-Scale Tunnel.

*F-22 model during tests in the Langley Spin Tunnel to determine
the size of emergency parachute required for spin recovery.*

F-22 model mounted for supersonic test in the Langley Unitary Tunnel.

Performance, stability, and control tests at supersonic speeds were accomplished in the Langley Unitary Plan Wind Tunnel in 1991.

A contribution from the Langley Spin Tunnel to the F-22 Program involved a relocation and assessment of the attachment point for the emergency spin recovery parachute. Evaluations of the original F-22 parachute truss structure and attachment point were accomplished in 1994; however, the attachment point had to be moved forward a distance of 3 ft on the full-scale aircraft to clear the exhaust plume of the vectored engine in afterburner. These tests were urgently requested by the Air Force and Lockheed to complete the installation on the flight-test aircraft before spring of 1999. The staff of the Langley Spin Tunnel provided timely and critical contributions, including the fabrication of a new parachute truss for the existing F-22 spin model, conducting free-spinning tests in the Spin Tunnel, and evaluating a broad range of store loadings and mass asymmetries.

Vertical-Tail Buffet

In collaboration with the F-22 System Program Office and the Air Force Research Lab at Wright-Patterson Air Force Base, Langley researcher Robert W. Moses conducted tail buffet tests in the Langley 16-Foot Transonic Dynamics Tunnel. These tests assessed the vertical-tail buffet environment and explored the potential application of advanced buffet alleviation systems. In these tests, the 13.5-scale model of the F-22 that had been used for tests in the Full-Scale Tunnel was refurbished with a new starboard flexible vertical-tail surface that included a rudder driven by a hydraulic actuator. A rigid, portside vertical tail was instrumented with pressure transducers for measurements of unsteady pressures.

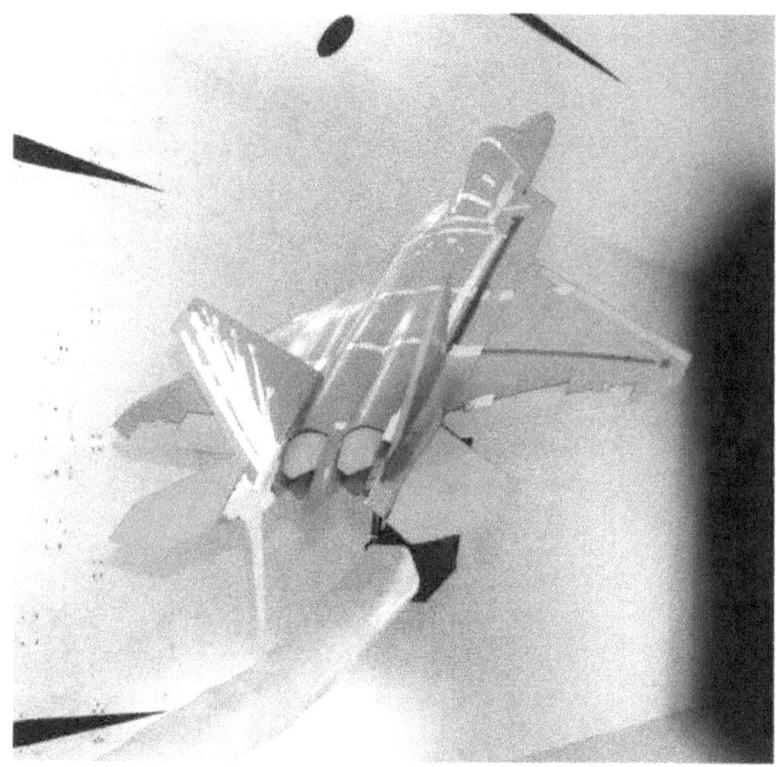

*F-22 model with instrumented vertical tails undergoes
buffet tests in the 16-Foot Transonic Dynamics Tunnel.*

Lockheed P-3 Orion

SPECIFICATIONS

Manufacturer
    Lockheed
Date in service
    July 1962
Type
    Antisubmarine warfare
Crew
    6 minimum
    11 normal
Engine
    Allison T56-A-14 turboprop

USERS

U.S. Navy, New Zealand,
Australia, Norway, Spain, Iran,
Japan, Netherlands, and
Portugal

DIMENSIONS

Wingspan . . . . 99.7 ft
Length . . . . 116.8 ft
Height . . . . 33.8 ft
Wing area . . . . 1,300 sq ft

WEIGHT

Empty . . . . 67,486 lb
Gross . . . . 139,760 lb

PERFORMANCE

Max speed . . . . .450 knot
Mission duration . . . . over
                                14 hr

HIGHLIGHTS OF RESEARCH BY LANGLEY FOR THE P-3

1. Tests in the Langley 16-Foot Transonic Dynamics Tunnel (TDT) identified a catastrophic propeller-whirl flutter that had caused two fatal accidents of the Electra civil transport on which the P-3 was based.

2. Engine mount modifications suggested by the results of the TDT tests cured the problem and the Electra (and the derivative P-3) safely completed service without further flutter issues.

In late 1958, Lockheed began production of a new four-engine turboprop civil transport named the Electra. About 170 were subsequently built and supplied mainly to U.S. and South American airlines. The Navy called for design proposals for a new aircraft to replace the aging P-2 Neptune for maritime patrol and antisubmarine warfare (ASW) in August 1957. To save cost and permit service introduction as quickly as possible, the Navy suggested that a variant of an existing aircraft or one in an advanced design stage would receive favorable consideration if suitable for the missions. Lockheed proposed a militarized version of the Electra, and in April 1958, the Navy awarded Lockheed a contract to develop the aircraft. The P-3V Orion entered the inventory in August 1962, and over 35 years later it remains the Navy's sole land-based antisubmarine warfare aircraft. It has gone through one designation change (P-3V to P-3) and three major models: P-3A, P-3B, and P-3C. The latter model is the only one now in active service. The last P-3 came off the production line at the Lockheed plant in April 1990.

Langley's contribution to the P-3 program involved powered-model tests of the predecessor Electra configuration in the Langley 16-Foot Transonic Dynamics Tunnel (TDT). During 1959 and 1960, two catastrophic Electra accidents occurred with evidence that the aircraft had experienced violent flutter and the wings had been torn from the aircraft.

Working under great pressure (about 130 Electras were in service at the time), a NASA and industry team conducted tests in the TDT along with analytical calculations. The driving mechanism behind the catastrophic structural failure was discovered to be propeller-whirl flutter. The phenomenon of propeller-whirl flutter involves a complex interaction of engine mount stiffness, gyroscopic torques of the engine and propeller combination, and the natural flutter frequency of the wing structure. The results of the TDT tests very accurately simulated the Electra scenarios. Based on these results, the engine mount structure on all Electras was strengthened, and the aircraft (and the derivative P-3) continued in service thereafter with great success and safety.

LANGLEY CONTRIBUTIONS TO THE P-3

The Lockheed Electra

In the fall of 1958, Lockheed began deliveries of a new four-engine turboprop Model L-188 civil transport to U.S. airlines. The sleek airliner, known as the Electra, set a new pace for luxury and speed for propeller transports. Over 170 Electras were built by Lockheed and delivered to U.S. and South American airlines.

In August 1957, the U.S. Navy called for design proposals for a new advanced aircraft for maritime patrol and antisubmarine warfare (ASW) to replace the aging P-2 Neptune. The Navy strongly suggested that a variant of an existing aircraft or one in the advanced design stage be used to save cost and permit rapid introduction into fleet service. Accordingly, Lockheed proposed a military version of the Electra. In April 1958, the Navy announced that the Electra derivative had won the competition. Known initially as the P-3V, the aircraft was redesignated in 1962 as the P-3 Orion. The Orion retained the wings, tail unit, basic fuselage structure, power plant, and many subsystems of the Electra, although its fuselage was about 7 ft. shorter than the Electra's.

Soon after introduction to the civil transport fleet, the Electra suffered two widely publicized fatal accidents with suspicious wreckage that raised concern over the structural integrity of the aircraft. On September 29, 1959, a Braniff Electra cruising near Buffalo,

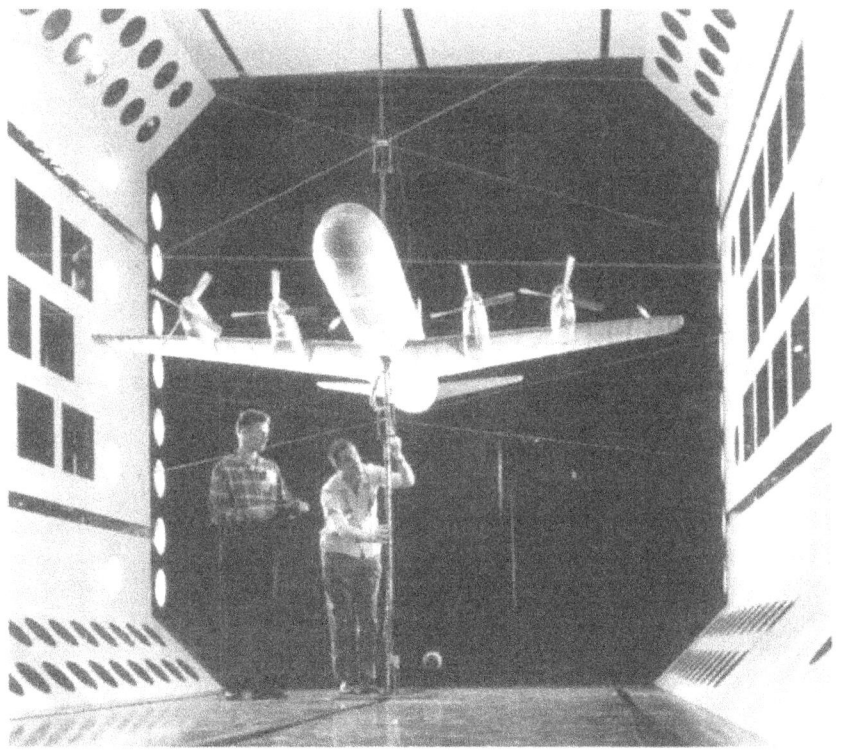

Powered model of the Lockheed Electra mounted in the Langley 16-Foot Transonic Dynamics Tunnel for flutter tests.

Texas, disintegrated without survivors. Investigation of the dispersal of aircraft wreckage revealed that the left wing had failed and separated from the aircraft in flight. On March 17, 1960, another Electra crashed in Indiana, with startling similarity to the Texas accident. Its right wing was found over 11,000 ft from the crash site, which indicated that it had also been torn from the aircraft. Over 130 Electras were operating in the civil fleet at the time, and authorities immediately reduced the cruise speed of the airliners while the investigation attempted to identify the cause of the fatal crashes.

Langley had recently completed the conversion of the Langley 19-Foot Pressure Tunnel into the world's first aeroelastic test tunnel, known as the Langley 16-Foot Transonic Dynamics Tunnel (TDT). The TDT had just been calibrated, and a 0.125-scale powered model of the Electra was quickly prepared for flutter tests. Senior researchers I. Edward Garrick, Philip Donely, Robert W. Boswinkle, and D. William Conner led the planning for the investigation. Frank T. Abbott and Robert M. Bennett led the experimental work in the TDT. Wilmer H. Reed, Jr. and Samuel R. Bland led analytical studies to model and predict the phenomenon.

Lockheed and Langley flutter experts were concerned that the propeller driven Electra may have exhibited the phenomenon known as propeller-whirl flutter, in which the stiffness of the engine mounts interacts with the gyroscopic torques produced by the engine and propeller combination. This interaction results in an unstable wobbling motion that could resonate with natural frequencies of the wing structure and could cause catastrophic flutter of the wing.

The industry and NASA team reduced the stiffness of the outboard engine mounts on the model and found that the wobbling motion indeed coupled with the natural flutter frequency of the wings. The fatal resonance could build up and tear the aircraft apart in 30 sec. Based on these results, the engine mounts on all Electra aircraft were strengthened, and the Electra and its derivative P-3 aircraft have since operated safely.

Lockheed S-3 Viking

SPECIFICATIONS

Manufacturer
  Lockheed
Date in service
  February 1974
Type
  Antisubmarine warfare,
  antisurface warfare, electronic
  warfare, and antimine warfare
Crew
  Four
Engine
  General Electric TF34-GE-400

USER

U.S. Navy

DIMENSIONS

Wingspan. . . . . 68.7 ft
Length . . . . 53.3 ft
Height . . . . 22.8 ft
Wing area . . . . .598 sq ft

WEIGHT

Empty . . . . 26,864 lb
Max takeoff . . . . 52,539 lb

PERFORMANCE

Max speed . . . . .450 knots
Range. . . . . 2,300 n mi

HIGHLIGHTS OF RESEARCH BY LANGLEY FOR THE S-3

1. Flutter clearance was obtained for the S-3 during tests in the Langley 16-Foot Transonic Dynamics Tunnel.

2. Spin and spin recovery characteristics of the S-3 were evaluated and found to be satisfactory during tests in the Langley 20-Foot Vertical Spin Tunnel.

Lockheed (now Lockheed Martin) S-3 Viking aircraft are tasked by carrier battle group commanders to provide antisubmarine warfare (ASW), antisurface warfare (ASUW), surface surveillance and intelligence collection, electronic warfare, mine warfare, coordinated search and rescue, and fleet support missions such as airborne refueling. The S-3B aircraft carries surface and subsurface search equipment with integrated target acquisition and sensor coordinating systems that can collect, process, interpret, and store ASW and ASUW sensor data. It has a direct attack capability with a variety of armament, including the Harpoon missile.

During operation Desert Storm, the versatility of the S-3 enabled it to serve a variety of roles. It flew hundreds of sea surveillance missions to enforce the economic blockade of Iraq and secure the seas surrounding the battle groups. Mine detection was an especially critical mission in the Persian Gulf, and the S-3 also served as the primary logistic transport for carriers. The S-3 had a direct combat role in which it electronically monitored active missile sites prior to air strikes. It also participated in the hunt for Scud missiles. Perhaps the most important contribution of the S-3 during the conflict was its airborne tanking capability. In this role, it employed a hose-and-drogue refueling system that resulted in more efficient refueling operations and more U.S. aircraft getting to their targets. In recognition of its broad applications, the designation of the S-3 Viking squadrons has been changed from Air Antisubmarine Squadron to Sea Control Squadron.

At the request of the Navy, Langley conducted wind-tunnel tests to provide flutter clearance and spin recovery evaluations for the S-3 prior to flight tests. The phenomenon of aileron "buzz" was encountered in the flutter test and eliminated, and flutter clearance was successful. Results of the spin tests indicated satisfactory characteristics, and no major modifications were made to the aircraft. Based on the data from the Langley tests, Lockheed and the Navy were able to proceed into flight-test development of the S-3 with increased confidence.

Langley facilities used in support of the development of the S-3 were the Langley 16-Foot Transonic Dynamics Tunnel and the Langley 20-Foot Vertical Spin Tunnel.

LANGLEY CONTRIBUTIONS TO THE S-3

Background

In 1964, the Navy announced a requirement for a new carrier-based antisubmarine warfare aircraft (designated VSX) to replace the aging Grumman S-2 Tracker. The new aircraft was to have twice the speed, range, and altitude capability of the S-2. Development of the S-3 Viking began in 1969 when Lockheed was awarded the VSX contract. Remarkably, only five years elapsed from the time of contract award until the first aircraft was delivered to a fleet squadron in 1974.

With the decline of the Soviet Union, and a subsequent reduction in the submarine threat to the U.S. fleet, the S-3 took on many new missions, including antisurface warfare, electronic surveillance, and airborne refueling.

Spin Tunnel Tests

In 1972, spin and spin recovery evaluations of the S-3 configuration were conducted in the Langley 20-Foot Vertical Spin Tunnel by Henry A. Lee and W. Louis White. Results of the tests indicated satisfactory characteristics would be expected of the S-3, and no modifications to the basic design were recommended. Although numerous stalling tests of the S-3 were conducted during the full-scale flight-test development program, intentional spins were not deemed necessary and were not attempted.

Operationally, the S-3 has not experienced stall-departure or spin problems.

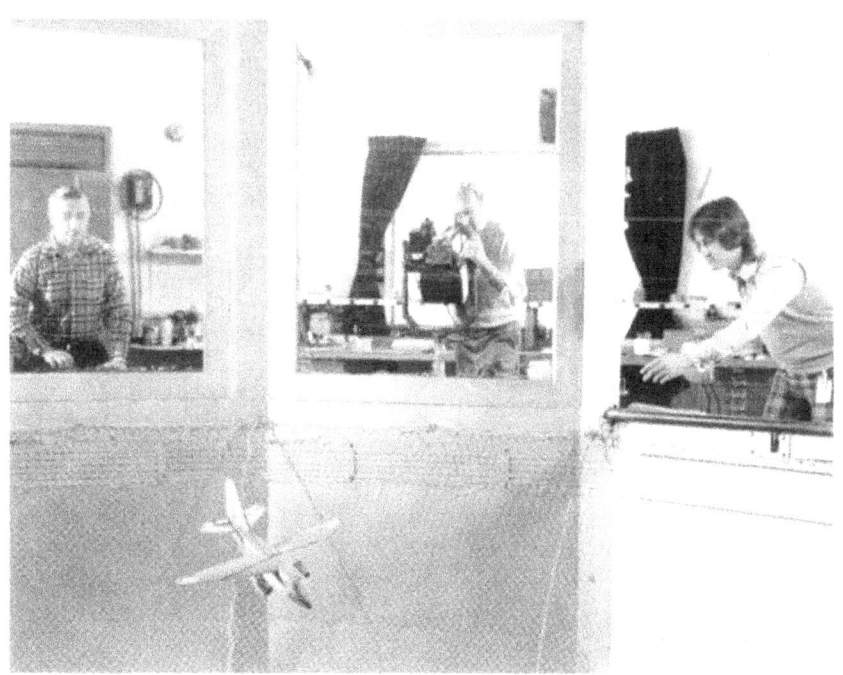

S-3 model launched into the vertically rising airstream in the Langley Spin Tunnel

Flutter Tests

Langley researchers Jean Gilman, Moses G. Farmer, and Charles L. Ruhlin conducted flutter tests of the S-3 in the Langley 16-Foot Transonic Dynamics Tunnel in late 1970 and early 1971. During the tests, the condition known as aileron "buzz" was encountered in which the S-3 ailerons exhibited very high frequency undamped oscillations. The problem was solved with a design modification to balance the ailerons, and flutter clearance for the complete configuration was demonstrated in mid-1971.

S-3 flutter model in the Langley 16-Foot Transonic Dynamics Tunnel.

McDonnell Douglas F-4 Phantom II

SPECIFICATIONS

Manufacturer
McDonnell Douglas
Date in service
December 29, 1960
Type
Fighter
Crew
Two
Engine
General Electric J79-GE-17

USERS

U.S. Navy, U.S. Marine Corps, U.S. Air Force, Egypt, Israel, Iran, Greece, Spain, Turkey, South Korea, West Germany, Australia, Japan, and Great Britain

DIMENSIONS

Wingspan . . . . 38.4 ft
Length . . . . 63.0 ft
Height . . . . 16.6 ft
Wing area . . . . .530 sq ft

WEIGHT

Empty . . . . 31,360 lb
Gross . . . . 58,000 lb

PERFORMANCE

Max speed . . . . . Mach number
of 2.0

HIGHLIGHTS OF RESEARCH BY LANGLEY FOR THE F-4

1. Studies in the Langley Unitary Plan Wind Tunnel revealed that the configuration exhibited lateral-directional stability problems at supersonic speeds, which resulted in McDonnell redesigning the wing and vertical tail.

2. Langley conducted studies of the stall departure and flat spin and identified causes of the poor stall-spin behavior of the F-4.

3. Langley, McDonnell Douglas, the Air Force, and the Navy investigated wing modifications to improve the maneuverability and handling characteristics, which resulted in incorporation of leading-edge slats on later versions of the aircraft.

4. Langley provided a solution to a severe nose buffet problem for the reconnaissance version.

5. Langley and Air Force braking performance tests conducted with an F-4 aircraft established runway grooving as an effective solution to hydroplaning accidents.

The McDonnell Douglas (now Boeing) F-4 was an exceptional fighter aircraft, which was used for air superiority, interdiction, and close air support. Originally developed by the Navy as a supersonic fleet defense fighter, the F-4 was also put into service by the Air Force and served a variety of roles in the Vietnam conflict. The final application of the F-4 by the U.S. was in the "Wild Weasel" role for suppressing enemy air defense systems. F-4 production ended in 1979 after over 5,000 had been built—more than 2,600 for the U.S. Air Force, about 1,200 for the U.S. Navy and U.S. Marine Corps, and the rest for friendly foreign nations. Later versions of the aircraft were in the U.S. Air Force inventory until December 1995. They are still flown by other nations.

NASA Langley Research Center contributions to the F-4 program began during the design stages when supersonic wind-tunnel tests at Langley identified stability issues that required redesign of the airframe, including adding dihedral (angle between horizon and wing panels from frontal perspective) to the outer wing panels and increasing the size of the vertical fin. Another contribution from Langley resulted in the development and incorporation of wing leading-edge slats for enhanced maneuverability and increased safety for high-angle-of-attack conditions.

Langley facilities used in the development and support of the F-4 program included the Langley Unitary Plan Wind Tunnel, the 30- by 60-Foot (Full-Scale) Tunnel, the 20-Foot Vertical Spin Tunnel, the 16-Foot Transonic Dynamics Tunnel, the 7- by 10-Foot High-Speed Tunnel, piloted simulators, and radio-controlled drop models.

LANGLEY CONTRIBUTIONS TO THE F-4

Early Configuration
Development

In 1954, the Navy selected McDonnell Aircraft to design and develop an all-weather supersonic fighter. The fighter, designated the F4H, was to be a fleet defense fighter that could take off from an aircraft carrier, have a cruise distance of 250 mi, intercept intruders, and then return to the carrier 3 hr after takeoff. The aircraft was to be armed with missiles and would not carry guns. It would operate as a high-speed (Mach number of 2), standoff missile launcher that would not engage in close-in combat.

In mid-1955 the full-scale engineering mock-up of the twin-engine aircraft featured a swept wing with no dihedral, and the horizontal tails drooped down at an angle of 15 deg. At the request of the Navy, tests began in the Langley Unitary Plan Wind Tunnel under the direction of Melvin M. Carmel to determine the supersonic performance and stability and control characteristics of the original configuration. Three separate entries in the Unitary Tunnel were made to evaluate the early F4H design.

Results of the first phase of tunnel tests indicated that the F4H exhibited serious deficiencies in lateral-directional stability characteristics at supersonic speeds, including unstable dihedral effect and marginal directional stability. To cure these problems, McDonnell introduced 12 deg of geometric dihedral into the outer wing panels (which were foldable for carrier operations) and increased the size of the vertical tail. Analysis had indicated that only 3 deg of geometric dihedral across the entire wing would solve the problem, but the same effect was achieved with less redesign and developmental effort by changing the outboard panels. Additional tests in other wind tunnels had indicated an undesirable pitch-up characteristic at transonic and low speeds, which was solved by adding chord to the outer wing panels (producing a leading-edge snag or "dog tooth") and by increasing the droop of the horizontal tail to 23 deg. The final phases of the F4H study in the Unitary Tunnel were directed to the integration of external stores at supersonic speeds.

Engineering mock-up of the original F4H Phantom without
dihedral in outer wings and 15-deg droop of the horizontal tail.

*F4H model after modifications with dihedral
in outer wings and increased vertical tail.*

After the development process, the F4H was placed into operations with the U.S. Navy and Marines under the new designation F-4 and set several speed and altitude records. The Air Force began to acquire F-4's in 1962, and this famous fighter became a mainstay in the air forces of friendly nations, with several variants produced by the McDonnell Douglas Corporation.

Maneuverability

During the first few years of the Vietnam conflict, the U.S. found itself engaging enemy aircraft such as the MiG-17 and MiG-19 that were relatively agile and could easily outmaneuver the heavier U.S. aircraft (F-4 and F-105) that had been designed without requirements for close dogfighting or close weapons such as a gun. Initial tactics used by U.S. pilots to try and turn with enemy aircraft had been relatively unsuccessful, and it had become apparent that missiles in use at that time were relatively unreliable at long ranges. Pilot training and revised tactics were ultimately employed to blunt the threat and use U.S. aircraft to an advantage, but the lack of maneuverability and a gun for close-in combat became issues for the Air Force. A new Air Force version known as the F-4E was equipped with a nose mounted M61 cannon, and additional deliveries to the Air Force began in October 1967.

McDonnell Douglas became interested in wing modifications for the F-4 that would improve buffet onset and increase lift and turning performance, while retaining satisfactory characteristics for approach and landing. Langley researcher Edward J. Ray led a cooperative Langley, McDonnell Douglas, and Air Force study of F-4 maneuver and buffet characteristics in the Langley 7- by 10-Foot High-Speed Tunnel in 1969. Candidate configurations included the use of wing leading-edge flaps, leading-edge camber, trailing-edge flaps, and other devices; however, the most effective modification was a two-position leading-edge slat. Two slats were mounted on the leading edge of each wing panel in place of the earlier leading-edge flap. The inner slat was fully retractable at high speeds, but the outer slat remained deployed in both the cruise and high-lift configurations. With the slats deployed, the F-4 could make tighter turns, and approach speeds were also reduced by a significant amount. Another benefit of this modification was a dramatic improvement in the lateral-directional handling characteristics and spin resistance at high angles of attack.

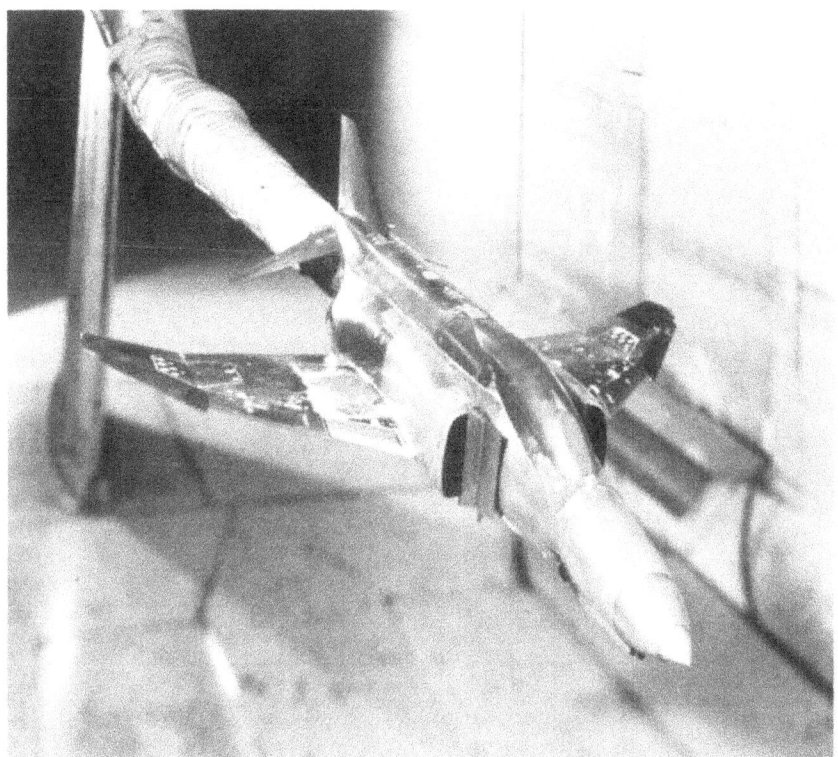

F-4E model in the Langley 7- by 10-Foot High-Speed Tunnel for evaluation of wing leading-edge modifications for enhanced maneuverability.

The slat configuration was evaluated during flight tests (known as Project Agile Eagle) of a modified F-4 test aircraft with extremely impressive results. The wing leading-edge slats were incorporated on all F-4E aircraft built during and after 1972. Later, the Navy received a slat equipped version of the aircraft known as the F-4S.

The F-4G Advanced Wild Weasel, which inherited most of the features of the F-4E including the wing slats, was one of the last versions of the F-4. Working in "hunter-killer" teams of two aircraft, such as F-4G and F-16C, the F-4G hunter could detect, identify, and locate enemy radar then direct weapons to ensure destruction or suppression of the radar. The technique was effectively used during Operation Desert Storm against enemy surface-to-air missile batteries.

High-Angle-of-Attack Stability Military strategists had procured the F-4 design as a standoff missile launcher with no requirement for strenuous maneuvering at high angles of attack. At that time, the need for close-in air combat maneuvering was thought to be obsolete, due to the advent of air-to-air missiles. Operations prior to Vietnam had stressed the supersonic mission for the F-4, and the initial safety record for the aircraft was extremely good.

The return of close-in air-to-air combat during Vietnam unfortunately exposed a deficiency in the flying characteristics of the F-4. During hard turns to engage or escape enemy aircraft, pilots began to fly the F-4 at high angles of attack where they experienced a marked deterioration in lateral-directional stability and control characteristics.

Inadvertent loss of lateral-directional control and spin entries occurred, with an alarming number of accidents and losses of crew and aircraft during training and combat. The Navy, Marine Corps, and Air Force suffered over 100 high-angle-of-attack and stall-spin accidents over the lifetime of the F-4 aircraft.

In 1967, representatives from the Air Force Aeronautical Systems Division met with Langley researchers Edward C. Polhamus and Joseph R. Chambers to discuss the growing concern over the high-angle-of-attack and stall-spin accidents (at that time, about 10 aircraft had been lost). Analysis of the problem by the Air Force was particularly difficult because the limited interest in high-angle-of-attack maneuvers had resulted in no wind-tunnel tests (other than spin tunnel tests) of the F-4 configuration for such conditions. Langley agreed to conduct diagnostic high-angle-of-attack wind-tunnel tests of the F-4. Tests in the Langley 12-Foot Low-Speed Tunnel and the Langley 30- by 60-Foot (Full-Scale) Tunnel thoroughly documented the aerodynamic factors that produced the loss of lateral-directional stability at high angles of attack.

Chambers and Sue B. Grafton conducted two free-flight model investigations of the F-4 for high-angle-of-attack conditions in the Full-Scale Tunnel. In the first study, flight tests of a model of the Air Force F-4C demonstrated an abrupt loss of directional stability (nose slice) near wing stall, and most of the flights ended with the model going out of control. Analysis of wind-tunnel data indicated that massive flow separation on the swept wing caused adverse flow fields at the tail, thereby degrading the stabilizing influence of the vertical fin at high angles of attack. No simple fixes could be found for the problem, short of a redesign of the wing.

When interest in wing leading-edge modifications for enhanced maneuverability became known, Langley conducted free-flight tests of a model of the later F-4E configuration. The test program, which was conducted by Edward Ray in the 7- by 10-Foot High-Speed Tunnel, was closely monitored. The results obtained from the Full-Scale Tunnel by Chambers and Grafton for high-angle-of-attack characteristics were fed back

Researcher Sue Grafton with the slatted F-4E model used for free-flight tests in the Langley Full-Scale Tunnel.

F-4 drop model mounted on the Langley helicopter drop rig for a spin entry test.

to Ray and others interested in maneuver performance. The results of the free-flight model test indicated that the incorporation of wing leading-edge slats markedly improved the high-angle-of-attack behavior of the F-4 by eliminating the severe nose slice tendency of the basic configuration.

Langley also conducted studies on the high-angle-of-attack characteristics of the F-4 with a radio-controlled model and a piloted simulator. The radio-controlled model was used by Charles E. Libbey to determine aircraft motions after loss of control at high angles of attack and to demonstrate the beneficial effects of the leading-edge slat modification. The simulator study of the F-4 by Frederick L. Moore was one of the first attempts to develop a simulator to study high-angle-of-attack behavior. The application of piloted simulators to this area is now a routine development tool.

The Flat Spin

Tests of an F-4 model by James S. Bowman, Jr. in the Langley 20-Foot Vertical Spin Tunnel in the early 1960's indicated that the F-4 configuration would exhibit several types of spins. In addition, the recovery characteristics from these various types of spins differed greatly. For example, one spin involved a relatively steep nose-down fuselage attitude with large oscillations in roll during the spin motion. Recovery from this spin was relatively easy with normal pilot control inputs. In contrast to this spin recovery, the F-4 also exhibited a relatively fast, flat spin in which the aircraft descended vertically in a rapid spin motion with an almost horizontal (flat) fuselage attitude. Recovery from the flat spin with normal aircraft controls was found to be impossible. During these spin tests, Langley determined that a 30-ft diameter emergency spin recovery parachute would be required for the test aircraft.

Spin tunnel tests are conducted by hand launching a model into a vertically rising airstream with an artificial spinning motion. The question remains as to whether the aircraft could enter such spinning conditions from conventional flight. Langley, therefore, conducts outdoor radio-controlled drop-model tests to evaluate this tendency. Drop tests were considered mandatory to confirm the potential for flat spins for the F-4. Fortunately, the results of the radio-controlled F-4 drop-model tests by Libbey indicated that the aircraft would have a dominant tendency to enter a steeper, recoverable spin rather than the flat, unrecoverable spin. In fact, the flat spin was only obtained once in over 20 aggressive attempts to spin the drop model.

The Navy and Air Force both conducted full-scale aircraft spin tests of the F-4, with aggressive control inputs to seek out the various spin modes. The Navy had expressed doubts over the existence of the flat spin at the beginning of their F-4 spin program. The agreement of the precursor Langley Spin Tunnel tests with the flight tests was remarkable. The steep, oscillatory spin mode was normally encountered and recovery was satisfactory as predicted by the Spin Tunnel tests. Unfortunately, the fast, flat unrecoverable spin was encountered in both test programs. The respective test aircraft were lost in crashes at Patuxent River Naval Air Station, Maryland, and Edwards Air Force Base, California, when the emergency spin parachute systems failed to operate properly. In the case of the Navy tests, the parachute failed to inflate properly, while in the Air Force tests, the mechanical system malfunctioned and released the parachute prematurely.

Although accident statistics for the F-4 indicated that the accidents involved about an equal number of approach-to-landing and up-and-away maneuvers, the Air Force and Navy expressed great interest in determining what features or factors of the F-4 were responsible for the unrecoverable flat spin mode. A request was made to Langley for assistance in determining these factors, and a contract was given to McDonnell Douglas to analytically study the flat spin.

During tests of the F-4 in the Full-Scale Tunnel, Chambers noted that a prospin yawing moment existed for the configuration at the attitudes associated with the fast, flat spin. Results from these tests also indicated that the prospin tendencies were produced by the aft end of the aircraft. An innovative, inexpensive test was used by Langley to identify the responsible aircraft components and the primary mechanism of the flat spin.

Chambers mounted a commercially available 1/48-scale plastic hobby model of the F-4 on a spindle assembly with a shaft through the top of the fuselage. The model was free to rotate as it would in a flat spin. When the hobby model was tested in the wind tunnel in a simulated flat spin attitude, researchers found that it would immediately spin up to the angular rates exhibited by the spin tunnel model and that the hobby model would do this to the right or left when released from a nonrotating condition. The researchers found that the driving mechanism for the flat spin tendency was adverse aerodynamic interactions between the drooped horizontal tail and the vertical fin. When the drooped tails were inverted (that is, the tails had 23 deg of positive geometric dihedral) the model would no longer accelerate into the spin condition and would stabilize without rotating.

The results of the hobby model tests were checked out in the Langley Spin Tunnel with the dynamically scaled F-4 model previously used for spin tests. The initial tests were conducted with the baseline F-4 configuration, and as expected, the model exhibited the fast flat spin. However, when the horizontal tails were inverted, the model would no longer spin flat, even though it was hand launched with conditions associated with the fast flat spin. Instead, the model would nose down into the steeper, recoverable spin. It was also found that increasing the deflection angle of the horizontal tail to 55 deg or greater eliminated the prospin aerodynamic interference effect for the basic configuration, which would no longer spin flat.

The results of these tests to identify the cause of the flat spin and other concepts identified by Langley (including an automatic spin prevention control system) were evaluated by the military. The recommended modifications to the F-4 fleet were considered too drastic at that time in the operational life of the aircraft, but the important adverse effect of the tail interference phenomenon was noted for subsequent aircraft development programs.

In the mid-1980's, the Navy approached Langley with reports that Marine pilots were recovering from flat spins by lowering the landing gear. The Navy requested spin tunnel tests and funded an F-4S model to determine the effects of operating the landing gear during the flat spin. Spin tunnel tests by Bowman showed that lowering the gear actually increased the spin rate in the flat spin! Based on these tests, the Navy did not incorporate lowering the gear as a spin recovery technique in the pilot handbook. Perhaps the most significant result from this test was that it alerted researchers to the potential adverse effects of components such as landing gear, gear doors, and weapons bay doors on spin recovery. The effects of such components had not been routinely tested in decades. This experience later proved very valuable during other programs.

The contributions and participation of Langley in the F-4 high-angle-of-attack and stall-spin efforts are also noteworthy because they formed the impetus within Langley to form a cohesive suite of test techniques designed to extract the maximum amount of information in this difficult and complex area. Following the F-4 program, Langley greatly increased commitments to this area and embarked on a research program with industry and DOD that has benefited the development of high-performance U.S. aircraft to this day.

Nose Vibrations

In late November 1965, the staff of the Langley 16-Foot Transonic Dynamics Tunnel (TDT) was urgently requested to assist the Air Force with a buffet problem that was experienced by the RF-4 reconnaissance version of the aircraft during combat operations in Vietnam. When pilots attempted to perform high-speed pre- and post-strike photographic reconnaissance missions, they found that the nose mounted cameras in the RF-4 were being literally shaken to pieces, seriously degrading the clarity of the photographic information and ruining the mechanical operations and operational lifetime of the cameras. Langley researchers Robert V. Doggett and Perry Hanson conducted wind-tunnel tests of the front fuselage of an RF-4 model in the TDT in early 1966 to measure fluctuating pressures in the area near the camera lens and housing. The results of the study indicated that aerodynamic flow separation on the lower forebody over the camera installation produced large vibratory loads that were the source of the problem. Doggett and Hanson, working with McDonnell Douglas and the Air Force, developed a modified camera ramp angle and revised camera enclosure fairing that eliminated the problem and was incorporated on later models of the RF-4. The fairing, which rounded the flat lower surface of the baseline nose, also resulted in increased internal nose volume and allowed larger cameras to be utilized.

The RF-4 reconnaissance version of the F-4 with camera fairing under the forward nose section.

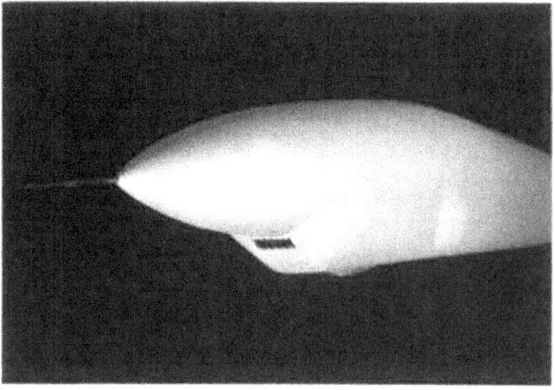

Simplified model of RF-4 nose tested in the Langley 16-Foot Transonic Dynamics Tunnel.

Runway Grooving

In the 1950's, Langley researchers at the Aircraft Landing Dynamics Facility focused their research on aircraft braking and directional control performance on wet runway surfaces. This research, led by Langley researchers Walter Horne and Thomas J. Yager, included ground handling tests of an F-4 test aircraft. Comparative tests of the effects of runway grooving on braking and handling characteristics provided clear demonstrations of the effectiveness of the grooving concept.

McDonnell Douglas F-15 Eagle

SPECIFICATIONS

Manufacturer
    McDonnell Douglas
Date in service
    November 1974
Type
    Air superiority, dual role
Crew
    One or two
Engine
    F-15C . . . Pratt & Whitney
    F100-PW-220

USERS

U.S. Air Force, Israel, Saudi
Arabia, and Japan

DIMENSIONS

Wingspan . . . . 42.8 ft
Length . . . . 63.8 ft
Height . . . . 18.5 ft
Wing area . . . . 608 sq ft

WEIGHT

Empty . . . . 40,000 lb
Typical combat . . . . . 68,000 lb

PERFORMANCE

Max speed . . . . above Mach
                    number of 2.0
Unrefueled . . . . 3,450 n mi
                    range

HIGHLIGHTS OF RESEARCH BY LANGLEY FOR THE F-15

1. At the request of the Department of Defense, Langley responded to a request for proposals with candidate fighter configurations of which one, known as LFAX-8, strongly influenced the winning design by McDonnell Douglas.

2. Langley conducted over 6,000 wind-tunnel test hours to collect data for the competitive source selection team and participated in the evaluation process.

3. Langley conducted studies to improve the subsonic cruise efficiency with emphasis on propulsion integration, which resulted in a substantial improvement in performance by the removal of ventral fins and modification of the vertical tails.

4. Langley conducted studies of high-angle-of-attack and spin characteristics to develop and demonstrate a high level of spin resistance.

5. Tests in the Langley 16-Foot Transonic Dynamics Tunnel identified potential flutter of the horizontal tails, which resulted in modifications of the tail geometry and subsequent flutter clearance.

6. Langley provided technology, expertise, and facility support for upgrades to the F-15 fleet on a continual basis including development of the F-15E and the F-15 short takeoff and landing and maneuver technology demonstrator (STOL/MTD).

One of the most illustrative examples of the benefit of NASA aeronautical research for military aircraft is the impact of NASA studies on the F-15 air superiority fighter. NASA studies ranged from conceptual configuration studies that employed aggressive technologies to developmental studies of aerodynamic performance, high-angle-of-attack and spin characteristics, flutter, and advanced derivative aircraft. Langley facilities used in the F-15 studies included the Langley 30- by 60-Foot (Full-Scale) Tunnel, the Langley 20-Foot Vertical Spin Tunnel, the Langley 4- by 4-Foot Supersonic Pressure Tunnel, the Langley Unitary Plan Wind Tunnel, the Langley 16-Foot Transonic Tunnel, the Langley 16-Foot Transonic Dynamics Tunnel, and the Langley 7- by 10-Foot High-Speed Tunnel.

In 1968, the Department of Defense (DOD) asked NASA to respond to the F-15 request for proposals (RFP) in a manner similar to the contractors to define the possible impact of advanced technology on the industry proposals. Under Langley's leadership, a NASA team of about 70 professionals colocated at Langley designed a series of advanced fighters that would meet the F-15 mission requirements. One of these configurations, LFAX-8, was of great interest to the McDonnell Douglas proposal team, which adopted many of the configuration features in the winning F-15 design.

Langley also participated in the source selection process by conducting over 6,000 test hours in eight wind tunnels to define the characteristics of competing F-15 designs and serving on the source selection evaluation board. During the development of the F-15, Langley provided continuing advice to the Air Force and provided a member to the F-15 Program Office. Langley's contributions to the F-15 in the areas of aerodynamics, propulsion integration, stability and control, aeroelasticity and flutter, and flight controls have helped the nation maintain superiority in advanced fighter aircraft.

LANGLEY CONTRIBUTIONS TO THE F-15

The NASA Fighter Study

Many of the basic design features of U.S. fighter aircraft have resulted from technology pioneered at the Langley Research Center. Some examples are the area rule, the variable-sweep wing, wing maneuver flaps, and efficient afterbody-nozzle integration. NASA continues to ensure that breakthrough concepts are quickly transferred to industry and DOD.

In 1967, Langley disseminated the results of in-house studies of a fighter configuration known as LFAX-8, which incorporated several features that would later be evident in the F-15 aircraft. Some of these features were

- Short propulsion package to minimize weight
- Engines placed forward for balance
- Horizontal ramp engine inlets for good performance at high angles of attack
- Horizontal tails located far aft on booms for increased stability and control
- Tailored twin-engine aft-end spacing and interfairing for efficient subsonic cruise conditions

Many industry design teams studied the database provided by Langley and these features were incorporated into many high-performance aircraft designs.

In 1968, the Department of Defense (DOD) requested that NASA respond to the F-15 request for proposals (RFP) in a manner similar to the industry contractors. The key person behind the NASA participation was Dr. John Foster, Director of the Defense Department Research and Engineering organization. He requested the participation for two reasons. First, Foster felt that NASA's aircraft designs for the F-15 mission would embody advanced technology and serve as the upper limit of technology for industry proposals. Second, NASA and its problem-solving expertise could minimize risks and problems later in the development program.

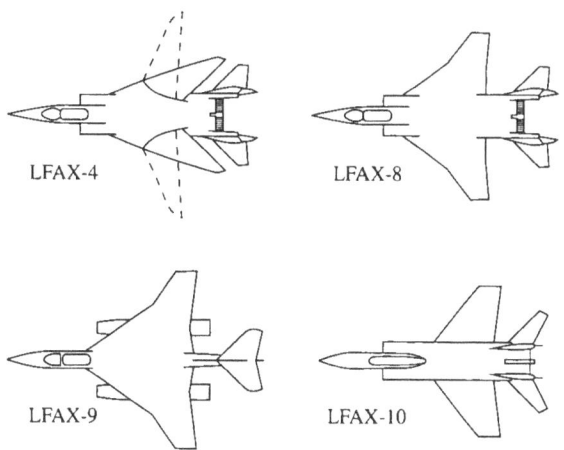

Fighter concepts developed by NASA for the F-15 mission requirements.

In response to Foster's request, NASA organized a design team of about 70 professionals from the Langley, Ames, Glenn, and Dryden Research Centers. The team was colocated at Langley under the leadership of Langley researcher William J. Alford, Jr. Roy V. Harris, Jr. and A. Warner Robins of Langley were members of a subteam responsible for configuration design and many other Langley researchers served on other subteams. The team spent about 4 months at Langley developing several fighter concepts directed at the F-15 mission requirements. The breadth of studies included analytical and wind-tunnel tests for the most promising configurations. Four fighter concepts were studied in great detail:

- LFAX-4—a variable-sweep configuration
- LFAX-8— a fixed-sweep version of LFAX-4
- LFAX-9—wing-mounted twin-engine configuration
- LFAX-10—similar in external shape to Soviet MiG-25 Foxbat

Industry design teams visited Langley during the efforts and were continually updated on the advantages, disadvantages, and technical maturity of the configurations. The NASA team also briefed high ranking DOD officials. The LFAX-4 and LFAX-8 embodied features that would subsequently be evident in the McDonnell Douglas F-15 and Northrop Grumman F-14 aircraft.

McDonnell Douglas was especially interested in the NASA fighter study as an extremely valuable adjunct to the company's design effort on the F-15. The LFAX-8 design made an indelible impression on the McDonnell Douglas design team, which embraced the fundamental layout of the NASA configuration. The cranked-wing design of the LFAX-8 had to be modified by McDonnell Douglas as the requirements for transonic maneuvering became more important. Another modification to the LFAX-8 involved the installation of a larger radar dish in the nose than the NASA team had allowed for in their design. The installation required a larger diameter nose cone, and although the NASA researchers deplored the increased supersonic drag caused by the larger nose, the final design incorporated the larger dish.

F-15 Source Selection

Following the NASA fighter study, 41 of the 70 NASA researchers became participants in the detailed evaluation of the industry proposals by McDonnell Douglas, Fairchild, and North American Rockwell. The studies and evaluation efforts, as well as developmental efforts at Langley, required over 6,000 hr in Langley facilities including the Unitary Plan Wind Tunnel, the 4- by 4-Foot Supersonic Pressure Tunnel, the 7- by 10-Foot High-Speed Tunnel, the 16-Foot Transonic Tunnel, the 16-Foot Transonic Dynamics Tunnel, the 20-Foot Vertical Spin Tunnel, and the 30- by 60-Foot (Full-Scale) Tunnel. During the development of the F-15, Langley personnel acted as consultants to the Air Force, and Langley provided a permanent member to the F-15 Program Office in Dayton, Ohio for improved liaison and communications.

Following the award of the F-15 contract by the Air Force to McDonnell Douglas in December 1969, Langley supported the development of the aircraft in several key areas.

Aerodynamic and Propulsion Integration Studies

Previous experiences with the F-111 and other advanced fighter concepts indicated that an extremely large portion of the subsonic cruise drag of modern twin-engine fighters is contributed by the aft end of the configuration (approaches 50 percent for some configurations). Langley researchers learned that careful tailoring of the engine interfairings and tail surfaces could prevent excessive aft-end drag. In the course of F-111

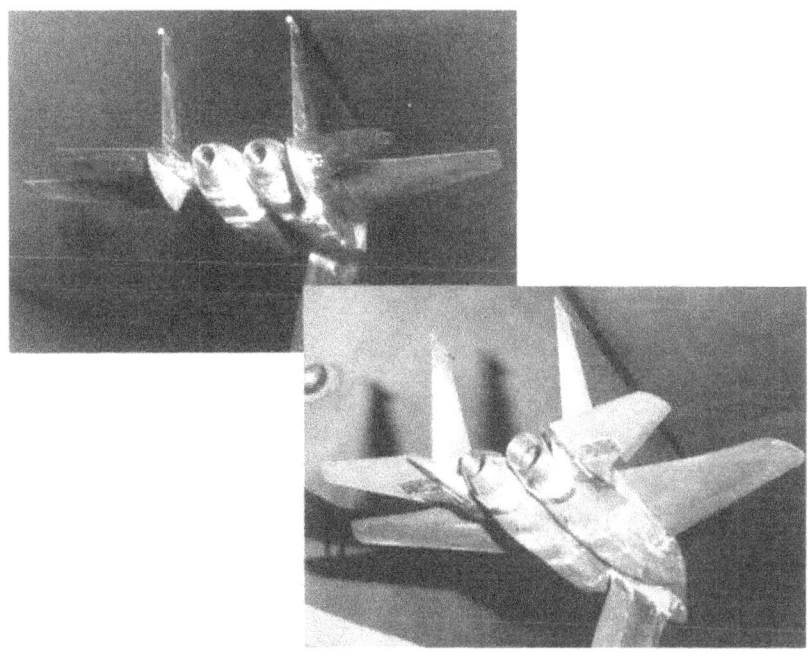

Aft-end configurations of F-15 models before (top) and after (bottom) Langley modifications such as removal of ventral fins, taller vertical tails, and nozzle interfairing.

development, Langley researchers in the 16-Foot Transonic Tunnel under Blake W. Corson, Jr. had developed test techniques and analysis methods to minimize this problem.

A Langley team led by Bobby L. Berrier conducted extensive tunnel tests on a 4.7-scale model of the F-15 for a critical assessment of aft-end drag on the baseline configuration, which at the time had large ventral fins below the aft fuselage. Mutual adverse aerodynamic interference effects were evaluated for the aft-end components, including the tails, tail booms, nozzles, and fuselage. After several tunnel entries, the Langley and McDonnell Douglas research team recommended configuration changes that significantly reduced the subsonic cruise drag of the aircraft. Specifically, the ventral fins were removed, the height of the vertical tails was increased to compensate for the resulting loss of directional stability, and nozzle interfairings (between the nozzles and also between the nozzles and tail booms) were added. Following briefings by Berrier to Air Force and DOD managers (including Dr. John Foster, Director of Defense Research and Engineering, and Robert Seamens, Secretary of the Air Force) of the test results and recommendations, the modifications (except for variable interfairings) were accepted for all production aircraft. The variable interfairings were actually built for flight-test evaluations, but were never flown after it was learned that the modified production aircraft met all critical mission requirements without the interfairing modification.

Upon completion of the drag clean-up studies, Berrier's team also provided the subsonic, transonic, and supersonic aerodynamic data package for the production aircraft. These data were obtained during multiple entries in the 16-Foot Transonic Tunnel and the 4- by 4-Foot Supersonic Pressure Tunnel. The final aerodynamic data package delivered by Langley included wind tunnel sting and distortion corrections and jet exhaust correction increments. Performance predictions of the final production aircraft were based on this data package.

In addition to studies of the production F-15, the Langley team conducted exploratory tests of the performance benefits of two-dimensional (2-D) nozzles and other advanced propulsion integration concepts. These research studies provided the largest U.S. database for advanced nozzle integration and served as a valuable foundation for development of the 2-D thrust-vectoring nozzles used by the F-15 short takeoff and landing and maneuver technology demonstrator (STOL/MTD) aircraft and the F-22 fighter.

Langley personnel conducted two other significant propulsion integration studies on the F-15. The first study was a wind tunnel to flight correlation of a highly instrumented powered model in the 16-Foot Transonic Tunnel and flight tests conducted at Dryden Flight Research Center. The second effort was a study of acoustically induced loads on the F-15 nozzles in the 16-Foot Transonic Tunnel and the Langley Acoustic and Dynamic Laboratory. The stimulus for this study, led by Langley researcher John M. Seiner, was the in-flight loss of external nozzle leaves ("turkey feathers") from operational F-15 aircraft as a result of structural fatigue. The cause and fix for this phenomena was identified during the Langley studies, but an alternate approach, at the expense of cruise drag, of simply removing the nozzle external leaves before they could fatigue and fall off was adopted by the Air Force.

High-Angle-Of-Attack and Spin Research

The Air Force and the Navy had expressed notable dismay over the relatively poor and unforgiving characteristics of the F-4 aircraft at high angles of attack. The F-4 exhibited a sudden directional divergence (nose slice) and other control-induced characteristics at high angles of attack that made the aircraft susceptible to loss of control and inadvertent spins. The two services lost a combined total of over 100 F-4's to accidents involving these characteristics during the operational life of the aircraft. Understandably, McDonnell Douglas approached the design of the F-15 with the intention of providing a high level of stability and spin resistance for the new air superiority fighter.

Langley technician Ronald White with one of two F-15 drop models used for research on spin-entry characteristics.

William P. Gilbert and James S. Bowman, Jr. led Langley research on the high-angle-of-attack and spin characteristics of the F-15. Facilities used for the effort included the Full-Scale Tunnel, the Spin Tunnel, and a helicopter-launched, radio-controlled drop model. McDonnell Douglas had done considerable homework as a result of the F-4 experience, and entered the F-15 development program with an appreciation of the stability and control features required.

In addition to careful layout of the airframe for satisfactory high-angle-of-attack stability of the F-15, McDonnell Douglas adopted a unique control augmentation system (CAS) in which movement of the control stick for lateral control resulted in deflections of the rudders, rather than the ailerons, at high angles of attack. With this approach, McDonnell Douglas restricted spin inducing aileron inputs while retaining adequate roll response. Initial analyses of this control concept by Langley researchers indicated that it would greatly enhance the spin resistance of the F-15. A demonstration of the spin resistance of the aircraft through the use of dynamically scaled free-flying models was still required. In addition, Langley was requested to identify the potential spin and recovery characteristics of the F-15 and the size of the parachute required for emergency spin recovery for a flight-test aircraft.

Results of tests of a 1/30-scale model in the Spin Tunnel indicated that the F-15 would exhibit developed spins if it exceeded the capability of the CAS to prevent spin entry. However, the ability of the aircraft control surfaces to recover the F-15 from spins was predicted to be very good. Meanwhile, tests in the Full-Scale Tunnel in 1971 on a 0.10-scale free-flying model and with a 0.13-scale model dropped from a helicopter indicated that the F-15 would be very stable at high-angle-of-attack conditions and that spin entry would be very difficult with the CAS active. Using the drop model, Langley researchers led by Charles E. Libbey defined a very limited range of control inputs and aircraft attitudes that enabled entry into a spin. In summary, the superior high-angle-of-attack behavior that had been anticipated by the McDonnell Douglas designers of the F-15 was validated and vividly demonstrated by the Langley dynamic-model tests.

In 1972, the Dryden Research Center assessed the impact of model size and Reynolds number on the results of the Langley tests. Dryden built and conducted drop tests with a 0.38-scale unpowered model, which was launched from a B-52. Results of the test program correlated well with the results obtained from the smaller model at Langley, and further substantiated the spin resistance of the F-15 configuration.

Flutter Clearance

Flutter clearance testing of the F-15 in the Langley 16-Foot Transonic Dynamics Tunnel required four tunnel entries. The tests were conducted in 1971 and were led by Langley researcher Moses G. Farmer. McDonnell Douglas provided a full-span F-15 model for the initial tests in the facility. Unfortunately, the F-15 was among a class of modern fighters that encountered tail flutter problems that were related to the pioneering use of composite materials in high-performance aircraft. The tunnel tests indicated that the original F-15 design would exhibit horizontal-tail flutter within its operational envelope. The early detection of this potentially catastrophic deficiency was a valuable contribution of Langley to the program. With the Langley results in hand, McDonnell Douglas engineers conducted flutter tests of the tail components in their own facilities. The problem was solved by redesigning the horizontal-tail geometry to control the aerodynamic center and mass balancing of the tail surface. By removing part of the inboard leading edge of the horizontal tail and adding additional balancing weight, McDonnell Douglas was able to clear the envelope of flutter for the development program. McDonnell

*Langley researcher Moses Farmer with F-15 model in preparation
for flutter tests in the 16-Foot Transonic Dynamics Tunnel.*

Douglas also removed a portion of wing near the wingtips to alleviate an objectionable
buffet characteristic at transonic maneuvering conditions; however, the requirements for
this modification came from flight-test evaluations.

*The F-15 Short Takeoff and
Landing and Maneuver
Technology Demonstrator*

In 1984, the Flight Dynamics Laboratory of the Air Force Aeronautical Systems Division awarded a contract to McDonnell Douglas for an advanced development STOL/
MTD experimental aircraft. The idea behind the program was to develop an aircraft that
could land and takeoff from sections of wet, bomb-damaged runway under bad weather
conditions and severe crosswinds without active ground-based navigational assistance.

The F-15 STOL/MTD aircraft was a direct adaptation of a configuration developed in
Langley sponsored programs conducted in the 1970's. Government and industry studies
of nonaxisymmetric two-dimensional (2-D) nozzles in the early 1970's had identified
significant payoffs for thrust-vectoring 2-D nozzle concepts. To better integrate the various 2-D nozzle technology programs and also to exchange data on this concept, a joint
NASA and DOD workshop was initiated and sponsored by Langley in September 1975.
To strengthen program integration and develop recommendations for further development of 2-D nozzles, an ad hoc interagency nonaxisymmetric nozzle working group was
formed at the workshop. Langley researcher Bobby Berrier was appointed as the NASA
representative to this group, which also included a member from the Air Force, the
Navy, and the Defense Advanced Research Projects Agency (DARPA) organizations.
The group advocated for a flight research program on the thrust-vectoring 2-D nozzle
concept.

The working group provided reviews and progress reports to senior management officials throughout the mid-1970's to early 1980's. The working group advocated for studies on three potential flight research vehicles. To avoid duplication, the Air Force
studied the F-111 vehicle, the Navy studied the YF-17/F-18 vehicle, and Langley studied the F-15 vehicle. In 1977, Langley initiated a system integration study of thrust-vectoring, thrust-reversing, 2-D nozzles on the F-15 with McDonnell Douglas. A

companion study of 2-D nozzle integration with the engine was initiated with Pratt and Whitney by the Glenn Research Center. The eventual F-15 STOL/MTD aircraft with thrust-vectoring, thrust-reversing, 2-D nozzles and a canard for trim of thrust-vectoring forces is a direct descendent of the configuration developed in these studies funded by NASA. In addition, multiple tests for both the F-15 generic and the F-15 STOL/MTD were conducted at the Langley 16-Foot Transonic Tunnel to study propulsion integration and power effect of thrust-vectoring, in-flight thrust-reversing (including methods for trim), 2-D nozzles on the F-15. As a result of its key role in developing the technology, Langley provided nearly all the features of the F-15 STOL/MTD aircraft.

The 16-Foot Transonic Tunnel team, led by Bobby Berrier, Francis J. Capone, Richard J. Re, Odis C. Pendergraft, Jr., Mary L. Mason, Laurence D. Leavitt, and others developed during the 1980's the most extensive design database in the world for low observable, thrust-vectoring, thrust-reversing (yaw and pitch), 2-D nozzles.

In recognition of Langley's leadership role in thrust-vectoring technology, the Air Force and McDonnell Douglas consulted frequently with the Langley staff, particularly with the propulsion integration experts under William P. Henderson at the 16-Foot Transonic Tunnel and the flight dynamics experts at the Full-Scale Tunnel. Free-flight tests conducted by Joseph L. Johnson, Jr.'s group in the Full-Scale Tunnel demonstrated that the aircraft would have excellent handling characteristics at high angles of attack. The most impressive research results in flight dynamics, however, occurred when the Langley researchers equipped the free-flight model with special 2-D nozzles that provided thrust vectoring in yaw as well as pitch. The superior control provided by the multiaxis vectoring was demonstrated when the model was easily flown at angles of attack up to about 85 deg without vertical-tail surfaces.

The first flight with the thrust-vectoring nozzles took place on May 16, 1989. The aircraft was transferred to Edwards Air Force Base for joint flight tests by the Air Force and McDonnell Douglas. The 2-D nozzles were first tested in flight on March 23, 1990. Test flights demonstrated a 25-percent reduction in takeoff roll, and the thrust-reversing feature made it possible for the F-15 to land on just 1,650 ft of runway (7,500 ft is

The F-15 STOL/MTD flying with canards and thrust-vectoring nozzles.

required for the standard F-15). In addition, thrust reversal was used during up-and-away flight to produce rapid decelerations—a useful feature for close-in air-to-air combat. During the flight program, the F-15 STOL/MTD made vectored takeoffs with rotation demonstrated at speeds as low as 42 mph. The program ended on August 15, 1991, after accomplishing all of the flight objectives.

Missiles

HIGHLIGHTS OF RESEARCH BY LANGLEY FOR MISSILES

1. From the 1950's Langley has worked to improve aerodynamic performance and roll characteristics of the Sidewinder missile, which resulted in rollerons on the tail fin.

2. Langley wind-tunnel studies assessed the effects of mounting Sidewinders on wingtips and pylons on fighter aircraft stability and maneuverability.

3. Langley participated in the development of the Sparrow missile in the early 1950's with flight tests to improve booster performance and supersonic wind-tunnel studies that resulted in changes to the nose and fin design.

4. Langley determined Sparrow missile carriage and separation loads for several aircraft.

5. Langley studied of the aerodynamics and relative drag of certain carriage configurations of the Advanced Medium Range Air-to-Air Missile.

6. Langley provided aerodynamic data from wind-tunnel tests during development of the Patriot missile.

7. Langley helped improve the stability problems of the early Hawk missile by proposing cutouts in the wings.

8. At the request of the Department of Defense, Langley assesses the impact of missiles and external stores on aircraft performance, spin recovery, and flutter.

As a result of unique supersonic wind tunnels and broad expertise across several technical disciplines, the NASA Langley Research Center has conducted extensive fundamental and applied research on missile configurations. Langley also conducts investigations to define the characteristics of specific missiles when requested by the Department of Defense (DOD). Langley typically addresses the aerodynamic performance, stability, and control characteristics of missile configurations; the aerodynamic phenomena associated with the carriage and release of missile shapes; and the effects of missiles and other external stores on the performance, stability and control, spin recovery, and aeroelasticity characteristics of the aircraft. Contributions have included the analysis and enhancement of missile shapes, the development and validation of advanced computational methods for missiles, and the development of unique wind-tunnel techniques for efficient and insightful tests. Langley has contributed to missile configurations for over 45 years in partnership with DOD and industry. These contributions are continuing for programs such as the Sidewinder, Sparrow, Advanced Medium Range Air-to-Air Missile (AMRAAM), Hawk, and Patriot missiles.

Langley facilities that supported research on advanced missile configurations included the 4- by 4-Foot Supersonic Pressure Tunnel, the Unitary Plan Wind Tunnel, the 16-Foot Transonic Tunnel, the 8-Foot Transonic Pressure Tunnel, and the Mach 6 High Reynolds Number Tunnel. Langley researchers also utilized the facilities at the NASA Wallops Flight Facility for missile testing.

LANGLEY CONTRIBUTIONS TO MISSILES

Background

In support of the Department of Defense (DOD), Langley Research Center is frequently requested to assist in the assessment and development of missiles. Langley provides support in the areas of aerodynamics, stability, and control for the isolated missile configuration, as well as in areas that might be affected by the carriage and release of the missile (aircraft performance, spin recovery, and flutter).

Langley became involved in the development of missile systems in the early 1950's, when a detachment of Langley researchers formed the core of what would become the NASA Wallops Flight Facility. These pioneering researchers used booster rockets to assess the aerodynamic characteristics of evolving air-to-air military missiles.

When large-scale supersonic wind tunnels were put into operation at Langley, special tests and test techniques were developed and validated for basic research for specific missile programs. Langley's leader in these early efforts was M. Leroy Spearman, who maintained a close working relationship with DOD and other agencies. In the 1960's and 1970's, Charles M. Jackson and Wallace C. Sawyer led the Langley missile program, which focused on a number of technical issues that were within the interests of the NASA program. The program included an expansion of the database for missile design (led by Jerry M. Allen and Adolphus B. Blair, Jr.), the development and validation of computational codes for missile aerodynamics (led by David S. Miller and Richard M. Wood), store carriage and separation (led by Robert L. Stallings, Jr.), and experimental techniques (led by William Corlett).

The vast majority of work in the Langley missile program involved research on representative missile shapes; however, very significant contributions were made to specific missile programs.

AIM-9 Sidewinder

In 1953, the Naval Ordnance Test Section requested assistance from Langley in the development of a short range, guided air-to-air missile for close-in combat. Known as the Sidewinder, the missile was to be heat seeking—that is, able to lock on to a target's heat producing sources, and featured forward canards for improved control.

At Wallops Island, Langley researchers conducted flight tests to improve the roll characteristics at supersonic speeds of the Sidewinder. During the tests, the missile was mounted atop a rocket test section: Langley developed special dampers, known as rollerons, for the rear control surfaces of the missile, which eliminated undesirable oscillatory roll tendencies. In 1956, Langley researchers studied how fighter aircraft would perform at supersonic speeds with various numbers of Sidewinder missiles stored in underwing locations and wingtip hard points. These studies were conducted in the Langley 4- by 4-Foot Supersonic Pressure Tunnel. The tests helped to establish limits on the external store capability of aircraft destined to carry the missile.

Through the years, programs have improved the maneuverability and range of the Sidewinder missile, and Langley has conducted numerous wind-tunnel test programs in the Langley Unitary Plan Wind Tunnel in cooperation with the Naval Weapons Center, China Lake, California. The most recent efforts in 1990, led by Blair and Allen, were conducted in the Unitary Plan Wind Tunnel and focused on the potential benefits of reduced tail span for enhanced supersonic maneuverability. The Langley and Navy team also examined a long range Sidewinder configuration, which incorporated an increased body diameter for more booster fuel.

Air Force crew loads a Sidewinder.

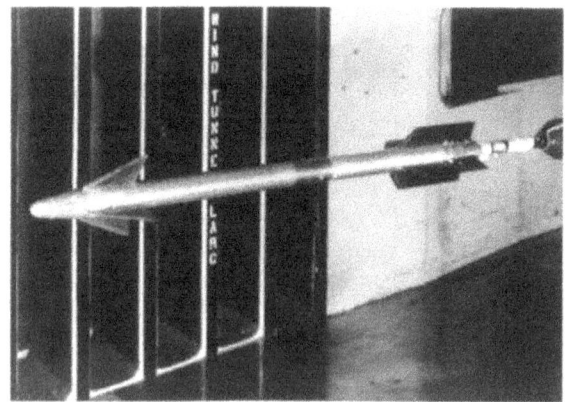

Sidewinder missile research model in the Langley Unitary Plan Tunnel.

AIM-7 Sparrow

In August 1948, Langley assisted the Navy and its contractors, the Sperry Corporation and the Douglas Corporation, in the development of the nation's first medium range guided air-to-air missile system. Langley constructed two uninstrumented development models of the missile, the XAAM-N-2 Sparrow I. The models were flown at the rocket range at Wallops to solve early booster problems. Refinements to the nose and fin sections of the missile were completed through subsonic and supersonic tests in Langley wind tunnels during the early 1950's.

Once the aerodynamics of the Sparrow configuration had been proven, the launching characteristics from positions under the wings, on the wingtips, and under the fuselage of the aircraft had to be determined for factors that might hinder aircraft and missile performance. With scale models, realistic simulations of missile launches were conducted during the mid-1950's in the Langley 7- by 10-Foot High-Speed Tunnel. Various fighter designs with the missiles carried externally (including the F-4) were tested at supersonic speeds in the Langley Unitary Plan Wind Tunnel to determine the aerodynamic loads on the missiles, as well as other external stores. Reliable and extremely accurate, the Sparrow later became the medium range workhorse for the Navy and Air Force.

Updates of the Sparrow missile system, such as the Sparrow III, also were tested in Langley wind tunnels. Initial tests of the advanced version began with an examination of the aerodynamic characteristics of the missile at supersonic speeds in the Unitary Plan Wind Tunnel in 1977. During the early to mid-1980's, further studies of the separation characteristics from simulated aircraft were conducted again in the Unitary Plan Wind Tunnel at speeds in excess of a Mach number of 2.

AIM-120 AMRAAM

By the mid-1980's, the military hoped to gradually replace the air-to-air medium range guided missile mainstay, the AIM-7 Sparrow with a more advanced weapons delivery system that would be carried by the next generation of fighters with supersonic cruise capability. The new missile, produced by McDonnell Douglas was designated the AIM-120 Advanced Medium Range Air-to-Air Missile (AMRAAM).

Initial aerodynamic studies of the AMRAAM at Langley focused on the missile itself, with scale models studied on a weapons palette. Extensive studies were conducted at supersonic speeds in the Langley Unitary Plan Wind Tunnel, first on a generic wing-body configuration and later on an advanced fighter configuration to determine the optimum external carriage locations. Researchers examined such factors as the amount of drag produced by missile arrangements and the amount of space required between missiles for effective weapons launch.

Today, the AMRAAM dominates the active radar missile field. Export orders for the AMRAAM totaled more than 7000, which is more than ten times the orders of its competitors combined.

Impact on Aircraft Characteristics

Since Langley is usually requested to support the development of high-performance aircraft, the Center is in an excellent position to assess the impact of missiles and other external stores on characteristics of the aircraft. Thus, assessments of the impact of missile configurations are normally included in performance evaluations in transonic and supersonic wind tunnels, studies of spin recovery characteristics in the Langley 20-Foot Vertical Spin Tunnel, and flutter clearance tests in Langley 16-Foot Transonic Dynamics Tunnel.

Northrop Grumman B-2 Spirit

SPECIFICATIONS

Manufacturer
  Northrop Grumman
Date in service
  December 17, 1993
Number built
  21 on order through year 2000
Type
  Multirole bomber
Crew
  Two (provisions for third)
Engine
  General Electric F118-GE-100
  nonafterburning turbofan

USER

U.S. Air Force

DIMENSIONS

Length . . . . 69.0 ft
Height . . . . 17.0 ft
Wing area . . . . 5,140 sq ft

WEIGHT

Empty . . . . 153,700 lb
Weapon load . . . . 40,000 lb
Max takeoff . . . . 376,000 lb

PERFORMANCE

Speed . . . . high subsonic
Range . . . . 6,000 n mi

HIGHLIGHTS OF RESEARCH BY LANGLEY FOR THE B-2

1. At the request of the Air Force, Langley specialists participated as members of B-2 technical review teams during design and early development stages.

2. Langley provided critical wind-tunnel tests with multiple entries in 5 Langley tunnels during the development program.

3. Areas addressed in the Langley tests included propulsion integration, high-angle-of-attack and low-speed stability and control, cruise performance, and spinning.

The B-2 Spirit's low observable, or stealth, characteristics give it the unique ability to penetrate an enemy's most sophisticated defenses and threaten its most valued, and heavily defended, targets. Northrop Grumman successfully met enormous technical challenges in blending the features and concepts required for low observability with the features required for high aerodynamic efficiency and large payload. In recognition of the significant accomplishments in the development of the B-2, Northrop Grumman was awarded the Collier Trophy in 1991. The NASA Langley Research Center provided key information for the design and development process by Northrop Grumman.

Some of the more obvious B-2 design challenges are apparent, including the efficient operation of highly integrated engine inlets; providing satisfactory stability, control, and handling characteristics for a flying-wing configuration without tail surfaces; and meeting mission performance specifications with high aerodynamic efficiency.

At the request of the Air Force, Langley researchers participated in preproduction technical reviews of the B-2 and later provided tests in several unique facilities at Langley during the development process. Langley facilities used for B-2 tests included the National Transonic Facility (NTF), the 16-Foot Transonic Tunnel, the 16-Foot Transonic Dynamics Tunnel, the 30- by 60-Foot (Full-Scale) Tunnel, and the 20-Foot Vertical Spin Tunnel.

LANGLEY CONTRIBUTIONS TO THE B-2

NASA supported Northrop Grumman in the aerodynamic design and development of the B-2 for over 15 years. Some of the more critical configuration development tests included the B-2 planform, the shielded upper-surface engine inlets, and the flight control system. At times, this effort approached 20 percent of the test operational schedule for some facilities.

B-2 model during tests in the National Transonic Facility (NTF) at Langley.

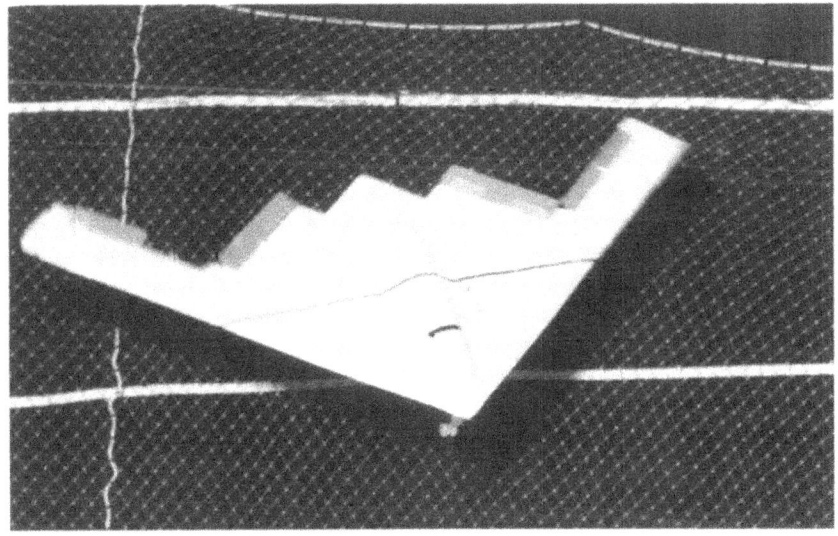

Free-spin and spin recovery tests of the B-2 in the Langley Spin Tunnel.

Rockwell B-1B Lancer

SPECIFICATIONS

Manufacturer
  North American Aircraft of
  Rockwell International
Date in service
  B-1B . . . June 1985
Type
  Bomber
Crew
  Four
Engine
  General Electric F101-GE-102
  with afterburner

USER

U.S. Air Force

DIMENSIONS

Wingspan. . . . . 137.2 ft
Length . . . . 145.8 ft
Height . . . . 33.6 ft
Wing area . . . . 1,946 sq ft

WEIGHT

Empty . . . . 186,234 lb
Gross . . . . 477,000 lb

PERFORMANCE

Max speed . . . . . Mach number
                          of 1.2
Range . . . . Intercontinental,
                        unrefueled

HIGHLIGHTS OF RESEARCH BY LANGLEY FOR THE B-1B

1. At the request of the Air Force, Langley participated in early studies and assessments of configurations that led to the B-1A and B-1B, including the Advanced Manned Strategic Aircraft (AMSA) Program.

2. The B-1 utilizes the variable-sweep wing concept that was matured by Langley.

3. Langley assessments of the high-angle-of-attack stability and control characteristics of the B-1A identified an undesirable longitudinal instability (pitch-up) that was eliminated by modifications to the flight control system.

4. Langley conducted several investigations of aft-end propulsion integration and the application of supercritical wing aerodynamics to reduce the cruise drag of the B-1A.

5. At the request of the Air Force, Langley conducted diagnostic wind-tunnel tests to analyze high loads on the engine nozzle flaps for the B-1B.

6. Flutter clearance tests and analysis of a unique wing oscillation problem were conducted at Langley.

The Rockwell (now Boeing) B-1B is a multirole, long range bomber, capable of flying intercontinental missions without refueling, then penetrating sophisticated enemy defenses. The B-1B's electronic jamming equipment, infrared countermeasures, radar location, and warning systems complement its low radar cross section and form an integrated defense system for the aircraft. Rockwell was awarded the Collier Trophy in 1976 in recognition of accomplishments for the B-1 development program.

Langley's association with the B-1 began with Air Force requests for active NASA involvement in several Air Force studies for advanced manned strategic bombers in the mid-1960's. Langley's involvement in advanced supersonic transports significantly enhanced the technical capabilities of the Air Force and Rockwell. As the B-1A Program progressed, the Langley staff conducted extensive tests and analysis of cruise-drag reduction, high-angle-of-attack behavior, and flutter. Following the cancellation of the B-1A Program and the initiation of the B-1B Program, Langley contributed to a joint NASA, DOD, and industry flight and wind-tunnel study to determine the cause of extremely high loads on the engine nozzle flaps for the B-1B aircraft.

Langley facilities involved in support for the B-1A and B-1B Programs included the Langley 30- by 60-Foot (Full-Scale) Tunnel, the 16-Foot Transonic Tunnel, the 8-Foot Transonic Pressure Tunnel, the 16-Foot Transonic Dynamics Tunnel, the 20-Foot Vertical Spin Tunnel, piloted simulators, and radio-controlled drop models.

LANGLEY CONTRIBUTIONS TO THE B-1B

Activities that led to the development of the B-1B bomber began in 1961, when the Air Force initiated studies under the Subsonic Low Altitude Bomber (SLAB) Program. The mission requirements and name of the bomber program continued to change during the early 1960's and became the Advanced Manned Strategic Aircraft (AMSA) Program in 1965. NASA's contributions in the definition of the FX Advanced Tactical Fighter (F-15) Program had not gone unnoticed by Air Force leaders involved in the AMSA Program. The close working relationships and mutual respect that existed between the Air Force and NASA are exemplified in excerpts from letters of request from the Air Force to NASA Headquarters for NASA participation in the AMSA effort.

From Lieutenant General Holzapple (Deputy Chief of Staff, Research and Development) to Dr. Mac Adams (NASA Associate Administrator for Advanced Research and Technology) on March 27, 1968:

> *It appears appropriate to have an independent review of the technical status of the (AMSA) air vehicle/engine configurations that have evolved. Informal discussions with Mr. Mark Nichols of NASA Langley indicate that NASA may be able to make a significant contribution in the form of evaluation and constructive criticism. Specifically, we have in mind verification of the latest preliminary designs and associated performance calculations and wind-tunnel results to ensure that our optimism is justified. In addition, such activity by NASA at this time would provide the appropriate background for NASA participation in any contractor source selection that might eventually come to pass.*

From General McConnell (Air Force Chief of Staff) to Mr. James Webb (NASA Administrator) on June 13, 1968:

> *The continued technical support that NASA has provided the Air Force in defining the FX Advanced Tactical Fighter has been most useful. I feel that our cooperative efforts have resulted in a very positive gain for the Air Force... Because of the value we place in NASA's expertise I would also ask you to consider supporting our AMSA effort. This should serve not only to improve our product but also to increase the confidence of others in our work.*

In response to these requests, Langley participated extensively in AMSA reviews, planning discussions, and exploratory wind-tunnel tests of candidate configurations. The expertise of the Langley staff in supersonic aerodynamics, and their active participation in the U.S. Supersonic Transport Program were especially valuable to the AMSA Team. The evolving requirements for the AMSA included challenging mission capabilities at supersonic speeds and low altitudes. As the studies began, it became obvious that application of the variable-sweep wing concept developed at Langley to the candidate bomber configurations was extremely beneficial.

THE B-1A PROGRAM

In June 1970, the Air Force awarded development contracts to North American Rockwell to build the B-1A bomber airframe and to the General Electric Corporation to provide the engines for the advanced bomber. Four B-1A prototypes were ordered for the development program. Langley immediately received a request from the Air Force to assist in the assessment and development of the B-1A.

High-Angle-of-Attack Research

The Air Force requested that Langley conduct the complete suite of tests for high-performance military aircraft, including wind-tunnel free-flight model tests, spin tests, and radio-controlled drop-model tests. Rockwell engineers were interested in the potential variation of handling characteristics with wing-sweep position and were anxious to uncover any problems, along with the solutions.

Free-flight model tests of the B-1A were conducted in the Langley 30- by 60-Foot (Full-Scale) Tunnel in 1972, by Langley researchers William A. Newsom, Jr. and Sue B. Grafton. Preliminary wind-tunnel tests of the model indicated that a severe longitudinal instability (pitch-up) would be encountered for angles of attack immediately above wing stall. This undesirable result was attributable to the relatively high position of the horizontal tail, which placed it in an adverse flow field at post-stall angles of attack. Subsequently, Langley radio-controlled drop-model tests by Charles E. Libbey vividly demonstrated the undesirable pitch-up characteristic. Piloted simulator studies were

Langley researcher William Newsom with the B-1A free-flight model.

B-1A model in free-flight tests in the Langley Full-Scale Tunnel in 1972.

B-1A drop model being prepared for helicopter drop test in 1974.

conducted at Langley by William D. Grantham and Langley pilot Perry L. Deal to evaluate the stall recovery characteristics of the B-1A under simulated mission conditions, such as hook-up and disengagement during air-to-air refueling. Grantham and his team of Langley and Air Force research pilots found that recovery from the pitch-up was very demanding and that a high potential for inadvertent secondary stalls (and pitch-ups) existed for the basic aircraft. The Langley, Rockwell, and Air Force team evaluated several control system concepts to artificially limit the angle of attack of the B-1A and thereby eliminate the post-stall instability from the operational envelope. Based in part on the Langley studies, an angle-of-attack limiter was subsequently incorporated into the B-1A flight control system.

Free-spin tests of a B-1A model were conducted in the Langley 20-Foot Vertical Spin Tunnel by James S. Bowman, Jr.; however, the angle-of-attack limiter system precluded serious concern of inadvertent spins with the B-1A.

Cruise Drag

At the request of the Air Force, Langley researchers analyzed and measured the cruise-drag characteristics of the B-1A during the design process and the development program.

At the request of the Air Force, Langley researchers William P. Henderson and Bobby L. Berrier participated in a special B-1A drag audit in the summer of 1972. The audit, which was requested by the B-1 Systems Program Office (SPO), consisted of a detailed assessment of the external and internal drag and installed thrust of the B-1A to evaluate the progress of the program and identify any high risk areas. The audit team consisted of 12 members from the Air Force and NASA.

Drag reduction studies for the B-1A were also conducted in the Langley 8-Foot Transonic Pressure Tunnel under the leadership of Theodore G. Ayers. The studies were directed at two objectives: the reduction of aerodynamic drag caused by the relatively large over-wing fairing and the potential application of supercritical wing aerodynamics. A cooperative Langley and Rockwell investigation of supercritical aerodynamics promised significant results; however, the Air Force directed Rockwell to turn their design information over to General Dynamics for possible application to the FB-111, and the B-1A application was not revisited.

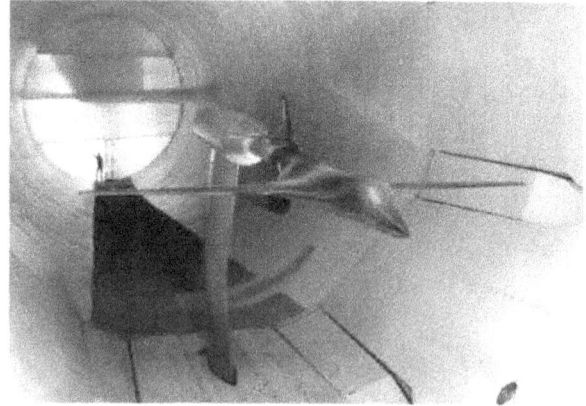

B-1A model in the Langley 16-Foot Transonic Tunnel.

Aircraft such as the B-1B that incorporate advanced configuration features such as variable-sweep wing and variable nozzle throat and exit areas for engines have detailed configuration features that can result in excessive drag at cruise conditions. For example, variable-sweep wings can have steps and gaps in the wing and glove juncture that can produce high drag. Variable exhaust nozzles when closed down to cruise settings often have large boat tail angles that result in high drag. Also, for twin-engine configurations with closely spaced exhaust nozzles, the large surface slopes in the gutter between the nozzles can cause flow separation and high drag.

To improve the drag characteristics of the B-1A, an investigation of several modifications on a 0.06-scale model was conducted in 1974 in the Langley 16-Foot Transonic Tunnel. Langley researchers Richard J. Re and David E. Reubush, with Rockwell and Air Force participants conducted tests to determine the effects of nozzle configuration, wing and glove fairings, fuselage underfairings, and other configuration modifications.

Flutter

The B-1A configuration underwent tests in the Langley 16-Foot Transonic Dynamics Tunnel (TDT) during several entries to determine flutter characteristics of the tail surfaces, obtain flutter clearance for the complete configuration, and investigate a unique outer wing oscillation phenomenon. Langley researchers Charles L. Ruhlin and Moses G. Farmer led these tests, which began in late 1972. Flutter clearance for the B-1A was obtained during this test program, and flight tests of the aircraft proceeded. During the flight tests at high altitude maneuvering conditions, an unusual aeroelastic phenomenon was encountered in which the outer wings of the aircraft exhibited a relatively low frequency, low amplitude undamped oscillation for certain values of wing sweep and flight conditions. Several entries in the TDT were conducted from 1973 to 1987 to further analyze the driving mechanism, which was identified as periodic shock-induced separation. The oscillations occurred near critical Mach number conditions for the wing airfoil and only at high positive angles of attack. The instability was demonstrated in the wind tunnel, however, at slightly different conditions than in flight.

On June 30, 1977, President Jimmy Carter announced that the United States would not proceed with the production of the B-1A bomber. However, flight tests of the four B-1A prototype aircraft continued until April 1981.

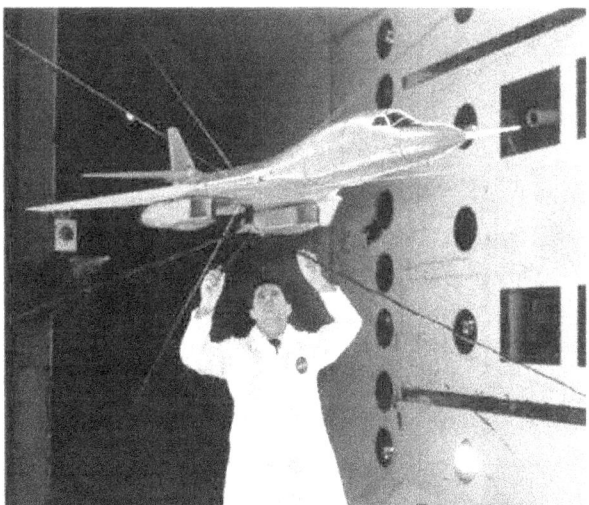

B-1A flutter model mounted in the Langley 16-Foot Transonic Dynamics Tunnel.

THE B-1B PROGRAM

As the B-1A Program was terminated, the DOD initiated a study through the Air Force Scientific Advisory Board to determine the need and direction for future strategic bombers. The results of the study concluded that a derivative of the B-1A, the B-1B, was the best candidate to fulfill the nation's needs within the envisioned mission requirements and the projected deployment date. Although the B-1B retained the same general geometrical shape of the B-1A, the shift in emphasis on penetration of highly defended targets resulted in modifications to the more crucial aircraft systems, especially defensive systems. The B-1B has a maximum speed of only half that of the B-1A, but it incorporates many more advanced concepts for enhanced survivability. Then known as the Long Range Combat Aircraft (LRCA), the B-1B was selected as the next strategic bomber and endorsed for production by President Ronald Reagan in October 1981.

Dynamic Loads on Engine Nozzle Flaps

The B-1B experienced excessive dynamic loads on the engine external nozzle flaps, which led to premature failures of the flap attachment system. In 1987, a joint Langley, Rockwell, and Air Force wind-tunnel and full-scale flight investigation of dynamic loads on engine nozzle flaps was conducted. A 0.06-percent full-span model of the B-1B with powered engine nacelles was tested in the Langley 16-Foot Transonic Tunnel under the leadership of Langley researchers John M. Seiner, Francis J. Capone, and Odis C. Pendergraft, Jr. The Langley researchers had previously participated in similar dynamic loads tests for the F-15. (See *Langley Contributions to the F-15.*) Seiner had led analysis of the principal mechanisms responsible for high loads on the external flaps that are caused by twin-plume supersonic resonance phenomena. The results of the B-1B study were similar to the previous F-15 analysis and contributed further understanding for improved design and analysis methods. Although the study team identified several solutions to the dynamic loads problem, the Air Force eventually decided to fix the problem by permanently removing the external flaps from the nozzles, which increased cruise drag, but reduced weight.

B-1B model in an inverted position for nozzle flap load investigation in the Langley 16-Foot Transonic Tunnel.

Rockwell-MBB X-31

SPECIFICATIONS

Manufacturer
  Rockwell
First Flight
  October 1990
Type
  Experimental demonstrator
  of combat utility of flight at
  extreme angles of attack
Crew
  One
Engine
  General Electric F404-GE-
  400

USERS

NASA (Langley and Dryden),
Rockwell, DARPA, U.S. Navy,
Deutshe Aerospace-MBB, and
German Ministry of Defense

DIMENSIONS

Wingspan . . . . 23.8 ft
Length . . . . 48.8 ft
Height . . . . 14.6 ft
Wing area . . . . .226 sq ft

WEIGHT

Takeoff . . . . 16,100 lb

PERFORMANCE

Max speed . . . . . Mach number
  of 1.28

HIGHLIGHTS OF RESEARCH BY LANGLEY FOR THE X-31

1. Langley contributed to exploratory studies of a fighter configuration designed to exploit high angles of attack in a precursor to the X-31 Program with Rockwell.

2. Working with Rockwell, Langley identified unacceptable characteristics of the initial design and successfully revised the configuration.

3. At the request of the Defense Advanced Research Projects Agency, Langley participated in the International X-31 Program and provided information on stability and control, control system effects, configuration effects, thrust-vectoring system, spin recovery, and recovery from out of control conditions.

4. During flight tests of the two X-31 research aircraft at NASA Dryden Flight Research Center, Langley provided facility support and technical consultation and analysis.

5. On two occasions, Langley provided timely solutions to critical X-31 deficiencies that had stopped the flight program.

The Rockwell (now Boeing) and Messerschmitt-Bolkow-Blohm (MBB) X-31 Enhanced Fighter Maneuverability (EFM) Demonstrator was designed to demonstrate the effectiveness of controlled maneuvers at extreme angles of attack during certain close-in air combat scenarios. The first International (U.S. and Federal Republic of (West) Germany) X-Plane Program showed the value of using thrust vectoring (redirecting engine exhaust flow) with advanced flight control systems to provide unprecedented levels of controlled flight to very high angles of attack. Whereas many previous fighters experienced loss of control in this regime, the X-31 was able to maneuver without fear of loss of control or inadvertent spins, which provided the pilot with new tactical options. The X-31, along with the NASA F-18 High Alpha Research Vehicle, was used in extensive flight tests at NASA Dryden Flight Research Center in the 1990's to provide the technologies and tactical evaluations to remove the high-angle-of-attack "barrier."

Langley became involved in the X-31 Program in 1984 in a cooperative research program with Rockwell to develop a fighter configuration capable of highly agile flight at extreme angles of attack. Free-flight model tests at Langley led to a major redesign of the Rockwell candidate configuration. When Rockwell, the Defense Advanced Research Projects Agency (DARPA), and the West Germans formed the X-31 Program, the staff at Langley was requested to participate in the configuration development. Langley researchers conducted extensive studies of the stability, control, and thrust-vectoring system of the vehicle. Langley remained active in the program as Dryden became the responsible test organization during the flight tests of two X-31 demonstrator aircraft. Flight tests began at Dryden in February 1992 and concluded in 1995.

During the flight evaluation tests at Dryden, Langley provided technical support and on two occasions provided rapid solutions to critical stability and control problems that had stopped the flight tests.

Langley support of the X-31 included tests in the 30- by 60-Foot (Full-Scale) Tunnel, the 20-Foot Vertical Spin Tunnel, the 12-Foot Low-Speed Tunnel, the 14- by 22-Foot Tunnel, the 16-Foot Transonic Tunnel, the Jet Exit Test Facility, a radio-controlled drop model, and piloted simulators.

LANGLEY CONTRIBUTIONS TO THE X-31

Background

Langley participated in the X-31 Enhanced Fighter Maneuverability (EFM) Program during four separate activities. From 1973 to 1984, Langley was active in the planning, testing, and analysis of the remotely piloted Highly Maneuverable Aircraft Technology (HiMAT) research vehicle. From 1984 to 1985, Langley cooperated in a program with Rockwell International to develop a representative fighter configuration that could demonstrate the advantages of exploiting high-angle-of-attack maneuvers during close-in air combat. From 1986 to 1991, Langley participated in the analysis and configuration development of the International (United States and Federal Republic of (West) Germany) X-31 Program. From 1991 to 1995, Langley supported the flight-test program, which was conducted at NASA Dryden Flight Research Center by the International Test Organization.

The HiMAT Program

Following the Vietnam conflict and renewed emphasis on close-in air-to-air combat, the U.S. military became interested in aircraft maneuverability. As a result, the requirement for high speeds, long considered the key factor in successful air combat, became a secondary objective. NASA initiated a joint program with the Air Force known as the Highly Maneuverable Aircraft Technology (HiMAT) Program. The staffs of the Langley, Ames, and Dryden Research Centers all participated in planning the HiMAT Program, with William P. Henderson serving as the technical lead and coordinator for Langley. The focus of the HiMAT Program was flight research and maneuverability demonstrations of a representative advanced configuration in the form of a remotely piloted subscale vehicle at Dryden. The goals of HiMAT included a 100-percent increase in aerodynamic efficiency over 1973 technology, and maneuverability that would allow a sustained 8-g turn at a Mach number of 0.9 and an altitude of 25,000 ft. The program ultimately achieved all goals.

The original HiMAT model with the thrust-vectoring wedge nozzle in the Langley 16-Foot Tunnel.

HiMAT during flight tests at Dryden.

In August 1974, Rockwell International was awarded a contract to construct a reduced scale model of the HiMAT design. Rockwell submitted a canard configuration with twin vertical tails on a highly swept wing and a high aspect ratio, pitch thrust-vectoring, wedge nozzle. The thrust-vectoring wedge nozzle was later replaced with a fixed, axi-symmetric nozzle to reduce program costs. The first flight of the HiMAT occurred on July 27, 1979, and research continued through January 1983. The success with the HiMAT configuration inspired Rockwell to examine the benefits of a derivative fighter that exploits high angles of attack for new tactical maneuvers during close-in air combat.

The SNAKE Program

In 1984, Rockwell proposed a cooperative program to Langley to assess and develop a Rockwell advanced design known as the Super Normal Attitude Kinetic Enhancement (SNAKE) configuration. Joseph R. Chambers and Joseph L. Johnson, Jr. determined that the proposal was in concert with many Langley research interests in high-angle-of-attack technology, and the cooperative program on the SNAKE configuration was begun. Langley researcher Mark A. Croom was assigned the role of lead engineer, and he began a decade of personal participation in the X-31 evolution and flight-test programs.

The initial SNAKE configuration bore a superficial resemblance to the earlier HiMAT design (canard and wing-mounted twin vertical tails); however, the new configuration was designed analytically with computers and a minimum amount of wind-tunnel tests. Unfortunately, Croom's aerodynamic tests of the initial SNAKE configuration in the Langley 30- by 60-Foot (Full-Scale) Tunnel indicated unacceptable stability and control characteristics. The configuration was unstable in pitch, roll, and yaw for all angles of attack.

Based on their extensive experience with stability and control characteristics of advanced fighters, Croom and Johnson provided the Rockwell team with several recommendations to cure the problems exhibited by the SNAKE configuration. The

*The original Rockwell SNAKE configuration with
downturned wingtips and thrust vectoring paddles.*

*The modified SNAKE configuration after the Langley tests with upturned wingtips, wing
leading-edge flaps, enlarged vertical tails, enlarged wing-fuselage strake, and nose strakes.*

*Modified SNAKE model flying at extreme angle of attack using
thrust vectoring in the Langley Full-Scale Tunnel.*

configuration modifications resulted in satisfactory characteristics, and the aerodynamic
deficiencies of the initial SNAKE design had been eliminated. Rockwell was grateful
for the guidance and innovation contributed by Langley in the evolution of the SNAKE
configuration.

In the early 1980's, an awareness of the benefits of thrust vectoring for dramatically
improved control at high angles of attack surfaced. In addition to studies of advanced
engine concepts with vectoring nozzles, interest arose over the use of simple thrust-
vectoring paddles in the engine exhaust to deflect the thrust for control augmentation.
As discussed in *Langley Contributions to the F-14*, the Navy, with Langley's assistance,
had taken the lead in this area with flight tests on an F-14 modified with single-axis
yaw-vectoring paddles. In addition, during a cooperative program with Rockwell led by
Langley researcher Bobby L. Berrier, Langley provided design data for multiaxis thrust-
vectoring paddle configurations using the Jet Exit Test Facility in 1985. Based on these
fundamental research studies, Rockwell incorporated multiaxis thrust-vectoring paddles
into the SNAKE configuration. Free-flight tests of the modified SNAKE model in the
Full-Scale Tunnel by Croom's team in 1985 provided an impressive display of the effectiveness of thrust vectoring at extreme angles of attack.

In West Germany, Dr. Wolfgang Herbst of Messerschmitt-Bolkow-Blohm (MBB)
aggressively touted the advantages of post-stall technology (PST) for increased

effectiveness during close-in air combat. Herbst's conclusions were based on wind-tunnel tests of a German advanced canard fighter configuration known as the TKF-90 and piloted simulator studies during which the application of simulated thrust vectoring resulted in rapid directional turns at high angles of attack had increased the turn rate by over 30 percent. Technical discussions between the Rockwell SNAKE Program managers and Herbst were initiated in 1983, and planning for a mutual program on PST ensued. Discussions with the Defense Advanced Research Projects Agency (DARPA) were very positive. When funding for collaborative international activities became available from the U.S. (the Nunn-Quayle research and development initiative in 1986) and West German governments, the technical expertise of Rockwell and MBB were joined under DARPA sponsorship in the X-31 Program. In view of Langley's extensive experience in high-angle-of-attack technology, unique test facilities, and contributions to the Rockwell SNAKE Program, DARPA requested in 1986 that Langley become a participant in the X-31 development program.

X-31 Configuration Evolution

The Rockwell and MBB X-31 design team merged their configuration candidates into a canard fighter powered by a single General Electric F404 engine with a single vertical tail. The initial design included an F-16 canopy for cost-saving purposes. Extensive tests of the initial X-31 configuration were carried out at Langley during 1987. These tests included static wind-tunnel tests and configuration component evaluations in the Langley 14- by 22-Foot High-Speed Tunnel, rotary-balance tests in the Langley 20-Foot Vertical Spin Tunnel to determine aerodynamic characteristics during spins, and dynamic force tests in the Langley Full-Scale Tunnel. Unfortunately, in 1988 the X-31 configuration was revised, and an F-18 canopy was incorporated. This change was regarded as significant, and a major portion of the previous wind-tunnel tests had to be repeated for the revised configuration.

Rotary-balance tests of the revised configuration were conducted in 1988, and spin tests and static and dynamic tests were completed in 1989 for the updated configuration. In 1989, a 0.19-scale model of the X-31 underwent extensive aerodynamic and free-flight tests in the Langley Full-Scale Tunnel. Results from these ground-based studies indicated that the X-31 might have marginal nose-down control at high angles of attack and that the configuration might exhibit severe, unstable lateral oscillations (wing rock) that would result in a violent, disorienting roll departure and an unrecoverable inverted stall condition. Fortunately, the results also indicated that a simple control law concept could prevent the aircraft from entering a spin. The awareness that such phenomena might exist for the full-scale aircraft enabled the X-31 design team to configure the flight control system for maximum effectiveness.

An exhaustive test, which included 498 paddle and nozzle configurations of the multiaxis thrust-vectoring system, was conducted by Langley researcher Francis J. Capone in the Jet Exit Test Facility during 1988. These data were used to select the final paddle and nozzle multiaxis thrust-vectoring configuration. These data were also critical to the design of the X-31 flight control system, since vectored thrust imposes large forces and moments in addition to the normal aerodynamic parameters.

A 0.27-scale drop model was used by Langley to evaluate the post-stall and out-of-control recovery characteristics of the configuration. The model, which weighed about 540 lb and included extensive instrumentation, was flown without an engine to assess the capabilities and characteristics of the basic airframe. The objective was to demonstrate that the X-31 would be agile and have satisfactory characteristics without the additional augmentation provided by thrust vectoring. The drop-model test identifies characteristics and large amplitude flight motions that cannot be assessed in

X-31 with F-16 canopy during tests in the Langley 14- by 22-Foot Low-Speed Tunnel.

Free-flight tests of the X-31 in the Langley Full-Scale Tunnel.

conventional wind or spin tunnels. In the X-31 Program, the technique proved to be invaluable as an early indicator of the highly unconventional behavior of the configuration. In particular, the violent roll departure indicated by tests of the free-flight model was encountered during the drop-model tests. Several control schemes were evaluated to eliminate this problem. In addition, the drop-model test technique provided solutions to barrier problems during the full-scale flight-test program.

X-31 Flight Demonstration Program

The first flight of the first X-31 aircraft occurred at Palmdale, California, on October 11, 1990, and the second aircraft made its first flight on January 19, 1991. During the initial phase of flight-test operations at the Rockwell facility at Palmdale, the two aircraft were flown on 108 test missions. On the test missions, the aircraft achieved thrust vectoring in flight and expanded the post-stall envelope to 40-deg angle of attack. Operations were then moved to Dryden in February 1992, at the request of DARPA.

At Dryden, the International Test Organization (ITO) expanded the flight envelope of the aircraft, including military utility evaluations that compared the X-31 to similarly equipped aircraft for maneuverability in simulated combat. The ITO, managed by DARPA, included NASA, the U.S. Navy, the U.S. Air Force, Rockwell Aerospace, the Federal Republic of Germany, and Deutsche Aerospace (formerly Messerschmitt-Bolkow-Blohm). The first NASA flight under the ITO took place in April 1992. As the X-31 full-scale aircraft flight tests began at Dryden, the Langley staff maintained a close support role for consultation and ground testing capability.

Two problems surfaced during the X-31 flight-test program, and both were considered significant enough to curtail flight tests until solutions were found. The first problem was encountered in the flight-test program when it became apparent that the pitch control effectiveness of the aircraft at post-stall conditions (particularly at aft center of gravity conditions) was marginal. Pilots reported that their ability to obtain positive, crisp, nose-down aircraft response was unsatisfactory and that increased control effectiveness was required if the X-31 was to be considered tactically responsive at high angles of attack. As part of the X-31 Team, Langley was requested to conduct wind-tunnel tests to explore options to provide the increased control at high angles of attack. Mark Croom and his team quickly responded and evaluated 16 configuration modifications to improve nose-down recovery capability in the Full-Scale Tunnel. Results of the

One of the X-31 aircraft in flight

Langley researcher Mark Croom (l) discusses the X-31 drop-model program with an X-31 program manager.

Mark Croom points to the aft-fuselage strakes defined by Langley tests and subsequently incorporated on the X-31 aircraft.

investigation recommended that a pair of 6- by 65-in. strakes be mounted along the fuselage afterbody to promote nose-down recovery. The Langley recommendations, which were given within a week of the test request, provided a timely solution to the problem. The aft-fuselage strakes were incorporated in the X-31, and the pilots reported that the nose-down control was significantly improved.

The second problem that occurred in the X-31 full-scale flight test was caused by large out of trim asymmetric yawing moments at high angles of attack. Shortly into the high-angle-of-attack, elevated-g phase of the envelope expansion, a departure from controlled flight occurred as the pilot was performing a maneuver at 60-deg angle of attack. Data analysis by the X-31 team indicated that a large asymmetric yawing moment, in excess of the available control power, had triggered the departure. In response to an urgent request for solutions, Croom and the Langley team conducted tests in the Langley Full-Scale Tunnel to design nose strakes that would minimize the problem. Once again, Langley responded rapidly with a strake configuration that permitted the flights to continue.

The X-31 Program logged an X-plane record of 524 flights in 52 months with 14 pilots from NASA, the U.S. Navy, the U.S. Marine Corps, the U.S. Air Force, the German Air Force, Rockwell International, and Deutsche Aerospace.

Evaluation of the X-31 as an enhanced fighter maneuverability demonstrator by the ITO concluded in early 1995.

Role of the X-31 in High-Angle-of-Attack Technology   The accompanying photograph shows three thrust-vectoring aircraft, each capable of flying at extreme angles of attack, cruising over the California desert in March 1994.

The F-18 HARV (top), the X-31 (middle), and the F-16 MATV (bottom) in flight.

The aircraft were flown in different programs and were developed independently. The NASA F-18 HARV was a test bed for aerodynamic and buffet data at high angles of attack to validate computer codes and wind-tunnel research. The X-31 was used to study the ability of thrust vectoring and advanced displays to enhance close-in air combat maneuvering. The F-16 Multiaxis Thrust Vectoring (MATV) aircraft was a demonstration of how thrust vectoring could be applied to an operational aircraft with an advanced engine that has a vectoring nozzle.

The Future—Investing in Freedom

As it looks to the future, the Langley Research Center intends to maintain a critical role in support of industries and government agencies that are responsible for the development of technology for the defense of the United States. However, there are several challenges in striving to maintain this critical role.

The level of Langley support for military research and development is dynamic and flexible. Langley participates as requested in response to the unpredictable international and domestic political and technical factors that influence our world as we enter a new millennium.

The reduction of U.S. military forces and plans for mergers or elimination of federal research laboratories complicate planning for aeronautical research programs. The numerous industrial mergers of the 1990's have consolidated the number of aircraft companies that request the results of NASA's aeronautics efforts. Diminishing budgets and personnel reductions have forced NASA to choose between space exploration and aeronautics when placing priorities, which has resulted in fewer resources for traditional military related research within NASA. Finally, maintaining world-class facilities in the face of increasingly competitive foreign facilities is a great challenge.

The end of the Cold War has diminished the intensity of the arms race. However, the alarming growth in international sales of sophisticated aircraft and arms to less developed countries and the development of advanced aircraft within nations such as the former Soviet Union and China continue to pose a technological challenge to the air supremacy of the United States.

The emergence of terrorist operations on a worldwide scale has changed the scope of the potential threat to the United States. This change in scope demands a new look at the adequacy of our military readiness and the requirements for new technology. For example, our experiences with uninhabited reconnaissance aircraft in Operation Desert Storm and European peacekeeping missions were highly successful. This success highlights the probability that uninhabited aircraft will increase in number in the future inventory of the United States. Uninhabited aircraft permit unprecedented deviations from existing design constraints and radical improvements within technical disciplines such as aerodynamics, structures, and flight controls.

Finally, the aging of the military aircraft fleet of the United States continues, and replacement aircraft will eventually be required. Nearly all military aircraft that participated in Operation Desert Storm, Kosovo, and other peacekeeping missions of the

1990's were over 20 years old. Some research activities that contributed to the development of these aircraft began over 50 years prior.

Future military aircraft development programs for the United States will begin from the extensive database of computer analyses, wind-tunnel tests, and flight-test results of past and present military aircraft. Information is added to this database by fundamental research conducted by Langley researchers. Very often this fundamental research points the way to the solution of any problem that may occur in an aircraft development program. Shrinking budgets and staff reductions further encourage the use and expansion of this database.

The staff at NASA Langley Research Center with their extensive aeronautical database and experience is ready to provide the research that is needed to maintain the air superiority of the United States. With adequate funding, Langley will continue in partnership with the Department of Defense, other government agencies, industry, and universities to develop the next generation of high-performance military aircraft.

Appendix

Wind Tunnels—The Tools of the Trade

NASA Langley Research Center's world-class stature in aeronautics is a direct result of the combination of the expertise and dedication of its staff and the unique test capabilities provided by its facilities. Throughout its long history, Langley has strived to anticipate requirements for new testing facilities, to conduct pilot testing of evolving facility concepts, and to provide updated test capabilities during new aircraft development programs. As a result of these efforts, Langley offers high quality, unique test capabilities in the areas of fundamental aerodynamics, aerodynamic performance, flight dynamics, aeroelasticity and flutter, spinning, structures and materials, impact dynamics, aircraft landing dynamics, and piloted simulators.

The discussion of Langley contributions in the text of this document illustrates the critical roles and responsibilities that were assigned to the Langley facilities during the development of these specific military aircraft. As might be expected, wind-tunnel facilities played a key role in the development process. Today, Langley operates wind tunnels that provide critical aerodynamic, flight dynamic, and aerothermodynamic data at speeds ranging from low subsonic conditions to hypersonic conditions. Donald Baal's and William Corliss' excellent review of NASA wind tunnels will provide the reader with extensive information on these wind-tunnel facilities (ref. 5).

The discussion of the military aircraft in this document reveals that certain Langley wind tunnels provided most of the data for the development of these military aircraft. These wind tunnels included the

- Langley 4- by 4-Foot Supersonic Pressure Tunnel
- Langley 7- by 10-Foot High-Speed Tunnel
- Langley 8-Foot Transonic Pressure Tunnel
- Langley 30-by 60-Foot (Full-Scale) Tunnel
- Langley 16-Foot Transonic Tunnel
- Langley 20-Foot Vertical Spin Tunnel
- Langley 16-Foot Transonic Dynamics Tunnel
- Langley Unitary Plan Wind Tunnel

New wind-tunnel test capabilities, such as those provided by the National Transonic Facility (NTF) at the Langley Research Center, and reductions in the NASA aeronautics operating budgets in the 1990's have resulted in the closure or transfer of the first four Langley wind-tunnel facilities on this list. The remaining wind tunnels on the list are still operating and providing valuable data for another generation of aircraft. Brief descriptions of the listed wind-tunnel facilities follow.

LANGLEY 4- BY 4-FOOT SUPERSONIC PRESSURE TUNNEL

Design work on the first large supersonic wind tunnel at Langley began in February 1945, with an effort to develop a 4- by 4-ft wind tunnel capable of providing test speeds up to a Mach number of 2. Delayed by labor strikes and other difficulties, the tunnel did not begin operations until May 1948. Initially powered by 6,000-hp motors (because of limited electrical power at the time), the facility was repowered in 1950 with 45,000-hp motors. Many historically significant military aircraft configurations and missiles were tested in the facility, including the Century Series fighters (F-100, F-102, F-104, and F-105), the TFX (precursor to the F-111), and the B-58 Hustler supersonic bomber.

The 4- by 4-Foot Supersonic Pressure Tunnel was dismantled in 1977, and its drive motors, cooling towers, and support facilities were used in the construction of the National Transonic Facility (NTF), which was built on the same site. NASA saved an estimated $20 million in construction costs by utilizing the components of the old tunnel to provide key elements for the new cryogenic NTF facility.

LANGLEY 7- BY 10-FOOT HIGH-SPEED TUNNEL

In 1945, Langley initiated operations in two 7- by 10-ft wind tunnels, which were built side by side in the same building. One of these tunnels, known as the Langley 300-MPH 7- by 10-Foot Tunnel, was used for basic and applied aerodynamic research at speeds up to 300 mph. The second tunnel, known as the Langley 7- by 10-Foot High-Speed Tunnel, was used for research at transonic test conditions.

The 7- by 10-Foot High-Speed Tunnel was the site of some of the most important Langley aerodynamic research for military aircraft. Researchers utilized the facility to conceive and develop breakthrough technical concepts such as the variable-sweep wing and vortex-lift strakes. It was also used to assess and support the development of most of the high-performance U.S. military fighters.

Regarded as one of the most productive research wind tunnels ever operated at Langley, the 7- by 10-Foot High-Speed Tunnel was closed and dismantled in 1993 because of limitations in operating budgets and workforce.

LANGLEY 8-FOOT TRANSONIC PRESSURE TUNNEL

Put into operation in 1953, the Langley 8-Foot Transonic Pressure Tunnel was used to produce some of the most important aeronautical breakthroughs for civil and military aircraft. Located on the original NACA property (East Area) at the Langley Air Force Base, the facility was a follow-on to the 8-Foot High-Speed Tunnel and incorporated the breakthrough slotted-wall concept for transonic testing. The new slotted tunnel was used to develop supercritical wing technology, winglets, and other innovative concepts that are now routinely used by military aircraft. Special test techniques, including the fluorescent oil flow visualization concept, were also developed and matured in the facility. The tunnel was utilized extensively for the support of military aircraft programs, including the assessment of competing designs, analysis of the performance capabilities of new configurations, and development of improvements and problem-solving concepts for evolving aircraft.

Following a multiyear research investigation of hybrid laminar flow control, the 8-Foot Transonic Pressure Tunnel was closed in 1995. The National Transonic Facility at the Langley Research Center and the Langley 16-Foot Transonic Tunnel have now taken on the transonic test workload.

LANGLEY 30- BY 60-FOOT TUNNEL

The design of the Langley 30- by 60-Foot (Full-Scale) Tunnel was initiated in 1929, and the tunnel was put into operation in 1931 for the intended purpose of obtaining full-scale aerodynamic data for the biplane configurations of the day. The tunnel contributed to military, commercial, and general aviation aircraft technology for over 64 years with a wide variety of test techniques including free-flight tests of remotely controlled, dynamically scaled models.

In 1985, the Langley Full-Scale Tunnel was named a National Historical Landmark. NASA closed this historic wind tunnel in 1995 as a cost-saving action. It was subsequently transferred to the Old Dominion University, which now operates the facility for a diverse customer base, including race car enthusiasts.

LANGLEY 16-FOOT TRANSONIC TUNNEL

Since becoming operational in 1941, the Langley 16-Foot Transonic Tunnel has undergone a series of improvements and upgrades. Initially capable of speeds up to a Mach number of 0.7, the tunnel can now test over a speed range of up to a Mach number of 1.29. The 16-Foot Transonic Tunnel is an atmospheric, closed-circuit tunnel with an octagonal test section that measures 15.5 ft across the flats. Twin 34-ft-diameter drive fans provide power, and the tunnel has an exceptionally low disturbance level through the transonic Mach number range.

From its earliest operations, the facility has specialized in propulsion-airframe integration issues, ranging from propeller research and engine cooling research during World War II to the design and integration of today's advanced multiaxis thrust-vectoring and reversing propulsion concepts. The breakthrough nonaxisymmetric (2-D) nozzle and thrust-vectoring nozzle technologies were primarily developed in this facility. The availability of high-pressure airlines, water cooling lines, and hydraulic lines permit this unique facility to conduct in-depth studies of powered advanced military configurations. Because of the large test section, which results in low blockage values for typical models, this facility has also been extensively utilized for transonic aerodynamic studies on a wide range of military aircraft.

The 16-Foot Transonic Tunnel is operational and continues to contribute to military aircraft development programs. This tunnel operates in conjunction with an auxiliary test

facility, known as the Jet Exit Test Facility, to provide design information for the development and maturation of innovative nozzle-propulsion system concepts and to support specific NASA and Department of Defense aircraft development programs.

Langley 20-Foot Vertical Spin Tunnel

Following the initial operations of a 15-ft-diameter spin tunnel in 1935, Langley designed and developed the Langley 20-Foot Vertical Spin Tunnel (Spin Tunnel) and initiated operations in 1941. The Langley Spin Tunnel is a closed-throat, annular return wind tunnel that operates at atmospheric conditions. The 12-sided test section is 20-ft across by 25-ft high. The test section velocity can be varied up to approximately 85 ft/sec. A 3-bladed, fixed-pitch fan powered by a 400-hp direct-current motor that is located above the test section produces test section airflow. This motor is equipped with a control system designed to allow rapid changes in fan speed, which results in rapid flow accelerations in the test section.

Dynamically scaled, free-flying models are used to investigate the spin and spin-recovery characteristics of aircraft configurations. To study spin characteristics, the model is hand launched with prerotation into the vertically rising air stream. The tunnel operator varies the tunnel speed so that the spinning model remains in equilibrium in front of video cameras for documentation of results. Direct observation of the test article is possible during tunnel operations via panoramic control room windows. The spin-recovery characteristics of aircraft are studied by using remote actuation of the model aerodynamic control surfaces. The size of emergency spin-recovery parachutes systems for flight test aircraft is also determined with specialized tests of scaled parachutes.

Free-spin data have been acquired in the Spin Tunnel and used by nearly all U.S. military fighter programs during and since World War II. The tunnel has now logged over 500 different aircraft studies. Nearly all U.S. attack and jet trainer programs have used Langley free-spin test data.

In addition to free-spin tests, the facility permits the measurement of aerodynamic forces and moments during spin conditions with a unique rotary-balance apparatus. In addition to providing aerodynamic input data for analyses and theoretical studies of spins, the rig has been used to provide electronically scanned pressures on models during simulated spin motions.

The Langley 20-Foot Vertical Spin Tunnel is the only facility in the United States currently configured for free-spin tests and determination of emergency spin-recovery parachute requirements for military aircraft. The facility continues its service to the nation for both military and civil spin studies. The majority of work, however, supports high priority military aircraft programs.

LANGLEY 16-FOOT TRANSONIC DYNAMICS TUNNEL

The NASA Langley 16-Foot Transonic Dynamics Tunnel (TDT) became fully operational in 1960. It is now internationally regarded as the leading wind tunnel for performing flutter tests of large aeroelastically scaled full-span models at transonic speeds. The TDT is the only facility in the world capable of studying a full range of aeroelastic phenomena at transonic speeds. The tunnel is used by the aircraft industry to clear new designs for safety from flutter, to evaluate solutions to aeroelastic problems, and to study aeroelastic phenomena at transonic speeds. The TDT is used to perform flutter clearance investigations and to investigate flutter trends and aeroelastic characteristics of fixed-wing and rotorcraft configurations. The TDT is also used to perform a variety of active controls tests, to determine the effect of ground-wind loads on launch vehicles, and to make steady and unsteady aerodynamic pressure measurements to support computational fluid dynamic (CFD) code development.

The TDT is a closed-circuit, continuous-flow wind tunnel capable of tests at stagnation pressures from near zero to atmospheric and over a Mach number range up to 1.2. The test section of the TDT is 16-ft square with cropped corners. One feature of the TDT that is particularly useful for aeroelastic testing is a group of bypass valves that connect the test section area to the opposite leg of the wind-tunnel circuit downstream of the drive fan motor. In the event of model instability, such as flutter, these quick-actuating valves are opened, which causes a rapid reduction in the test section Mach number and dynamic pressure, which may result in stabilizing the model. Other features that make the TDT uniquely suited for aeroelastic testing include high visibility of the model from the control room, a highly sophisticated data acquisition system, flow oscillation vanes upstream of the test section that can be used to generate sinusoidal gusts, a variety of model mounting and suspension systems such as cantilever sidewall mounts for component models and a 2-cable-suspension system for full-span free-flying models, safety screens that protect the tunnel fan blades from debris in case of a model failure, and state-of-the-art instrumentation and test equipment.

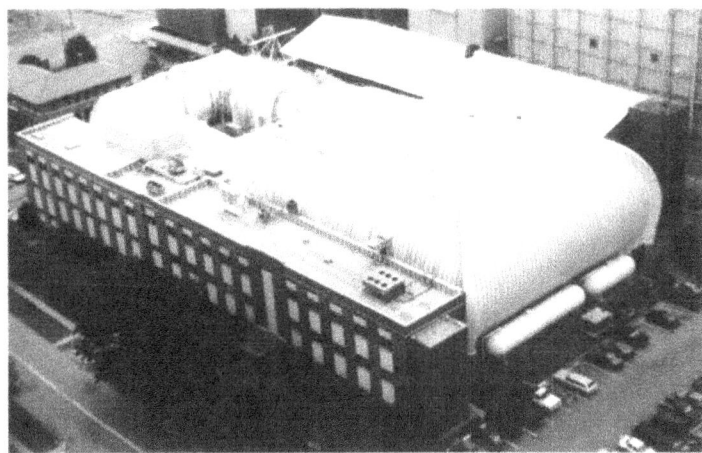

Tests can be performed in the TDT with air as the test medium; however, the most distinguishing feature of the tunnel is the use of a heavy gas, presently R-134a refrigerant, as the primary test medium. R-134a is about four times as dense as air, and yet sound travels half the speed in R-134a as it does in air. These properties of higher density and lower sonic speed have beneficial effects on the design, fabrication, and testing of aeroelastically scaled wind-tunnel models that must accurately represent full-scale counterparts. Physically larger models may be built, thereby simplifying the model fabrication process. The scaled natural frequencies of these larger models are lower, resulting in lower flutter frequencies, thereby reducing the risk of model destruction during flutter. Other advantages resulting from the use of a heavy gas are a nearly three-fold increase in Reynolds number and a lower tunnel-drive horsepower requirement.

The TDT has played a key role in the development process for virtually all military aircraft in the U.S. inventory, including transports and high-performance fighters. The facility is extremely active and supports civil and military programs.

LANGLEY UNITARY PLAN WIND TUNNEL

The Langley Unitary Plan Wind Tunnel has been in continuous operation since construction was completed in 1955. Congressional approval for the tunnel was provided by the Unitary Wind Tunnel Plan Act of 1949, which stated that the objective was to "Promote the national defense by authorizing a unitary plan for construction of transonic and supersonic wind-tunnel facilities..."

Developmental tests of virtually every supersonic military aircraft, missile, and spacecraft in the current U.S. inventory were performed in the Unitary Tunnel. In addition, methods for predicting supersonic aerodynamic performance have been developed through basic experimental fluid mechanics research conducted in the Unitary Tunnel. The Unitary Tunnel is a closed-circuit pressure tunnel with two 4-ft by 4-ft by 7-ft test sections, and a Mach number capability ranging from 1.47 to 4.63.

The Langley Unitary Plan Wind Tunnel participates in advanced military aircraft and missile research.

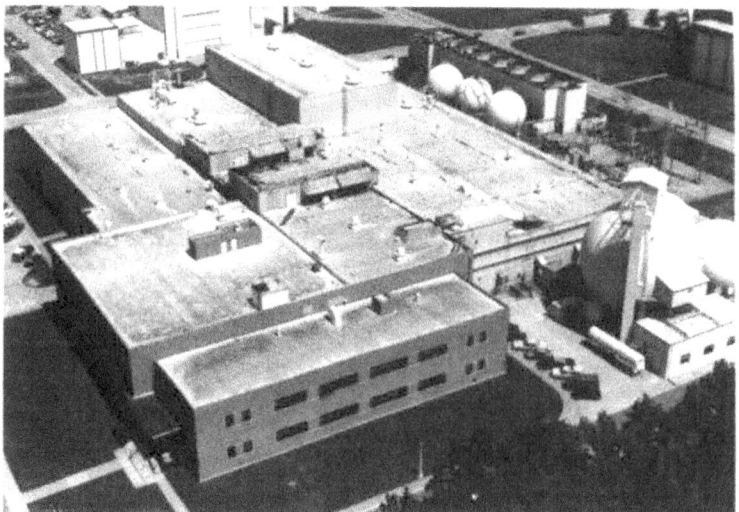

Bibliography

BAI Exdrone BQM-147A

1. Willmott, Jay: Memorandum from Vice President, BAI Aerosystems, Inc. to author. June 29, 1999.

2. Yip, Long; Fratello, Dave; Robelen, Dave; and Makowiec, George: Wind Tunnel and Flight Tests of a Delta-Wing Remotely Piloted Vehicle. *J. Aircr.*, vol. 28, no. 11, Nov. 1991, pp. 728–734.

3. UAVs Pressed into Action to Fill Intelligence Void. *Aviat. Week & Space Technol.*, vol. 135, no. 7, Aug. 19, 1991, pp. 59–60.

4. BAI Aerosystems, Inc. Unmanned Aerial Vehicles, Support Equipment, and Services. http://www.baiaerosystems.com Accessed Mar. 2000.

Boeing AV-8 Harrier

1. Myles, Bruce: *Jump Jet—The Revolutionary V/STOL Fighter*, Second ed. Brassey's Defence Publ., 1986.

2. Smith, Charles C., Jr.: *Flight Tests of a 1/6-Scale Model of the Hawker P.1127 Jet VTOL Airplane*. NASA TM SX-531, U.S. Air Force, 1961.

3. Chambers, Mark: Developing the World's First Jump-Jet. *Air Combat*, vol. 24, no. 3, May/June 1996, pp. 4–11.

4. Chambers, Mark: Pushing the Envelope. *Air Combat*, vol. 23, no. 6, December 1995, pp. 4–11.

5. Carter, A. W.: *Effects of Jet Exhaust Location on Longitudinal Aerodynamic Characteristics of a Jet V/STOL Model*. NASA TN D-5333, 1969.

6. Margason, R. J.; Vogler, R. D.; and Winston, M. M.: *Wind-Tunnel Investigation at Low Speeds of a Model of the Kestrel (XV-6A) Vectored-Thrust V/STOL Airplane*. NASA TN D-6826, 1972.

7. Siuru, William D., Jr.: *British Aerospace and McDonnell Douglas Harrier AV-8A/B*. Aero Publ. Inc., 1985.

8. Culpepper, R. G.; and Murphy, R. D.: *Unique Facility for V/STOL Aircraft Hover Testing*. NASA TP-1473, 1979.

9. Campbell, John Paul: *Vertical Takeoff and Landing Aircraft*. MacMillan Co., 1962.

10. Lacey, T. R.; Johnson, D. B.; and Voda, J. J.: Powered Wind Tunnel Testing of the AV-8B—A Straightforward Approach Pays Off. AIAA-79-0333, Jan. 1979.

11. Morello, S. A.; Person, L. H., Jr.; Shanks, R.E.; and Culpepper, R. G.: *A Flight Evaluation of a Vectored Thrust Jet V/STOL Airplane During Simulated Instrument Approaches Using the Kestrel (XV-6A) Airplane*. NASA TN D-6791. 1972.

Boeing C-17 Globemaster III

1. Chambers, Mark: New C-17 Aircraft Blows Into Langley. *Researcher News.* Langley Research Center, May 31, 1996.

2. Campbell, John P.; and Johnson, Joseph L., Jr.: *Wind-Tunnel Investigation of an External-Flow Jet-Augmented Slotted Flap Suitable for Application to Airplanes With Pod-Mounted Jet Engines.* NACA TN-3898, 1956.

3. Rivera, Jose A.; and Florance, James R.: Contributions of Transonic Dynamics Tunnel Testing to Airplane Flutter Clearance. AIAA-2300-1768, 2000.

4. Pearson, Robert M.; Giesing, Joseph P.; Nomura, James K.; and Ruhlin, Charles L.: *Transonic Flutter-Model Study of a Multijet Airplane Wing With Winglets Including Considerations of Separated Flow and Aeroelastic Deformation Effects.* NASA TM-87753, 1986.

Boeing F-15 Eagle

1. Harris, Roy V.; et al: Summary of Langley Contributions to the F-14, F-15, F-16, and F-18 Fighter Aircraft. Memorandum to the Director From Chief, High-Speed Aerodynamics Division. NASA Internal Memo, Aug. 26, 1981.

2. Alford, William J., Jr., compiler: *NASA Fighter Aircraft Study in Support of the USAF F-15 Tactical Fighter: Volume I—Summary, Configurations, and Mission Analysis.* NASA LWP-707A, 1969.

3. Staff of the NASA-LRC F-15 Study Group: *Static Aerodynamic Characteristics and Exhaust-Nozzle/Afterbody Characteristics of the McDonnell Douglas F-15 Airplane (199-1B) Design.* NASA LWP-811, 1969.

4. Staff of the Large Supersonic Tunnels Branch: *Aerodynamic Characteristics at Mach Numbers From 1.60 to 2.86 of a 0.047-Scale Model of the McDonnell Douglas F-15A Airplane (199-1B).* NASA LWP-774, 1969.

5. Staff of the NASA-LRC F-15 Study Group: *Static Aerodynamic Characteristics and Exhaust-Nozzle/Afterbody Characteristics of the North American Rockwell F-15 Airplane (D490-3B) Design.* NASA LWP-805, 1969.

6. Staff of the NASA-LRC F-15 Study Group: *Static Aerodynamic Characteristics and Exhaust-Nozzle/Afterbody Characteristics of the Fairchild Hiller F-15 Airplane Design.* NASA LWP-815, 1969.

7. Ray, Edward J.; and Fox, Charles H., Jr.: *Subsonic Drag Analysis of a 7.5-Percent F-15 Model.* NASA LWP-1014, 1971.

8. Berrier, B. L.; and Maiden, D. L.: *Effect of Nozzle-Exhaust Flow on the Longitudinal Aerodynamic Characteristics of a Fixed-Wing, Twin-Jet Fighter Airplane Model.* NASA TM X-2389, 1971.

9. Maiden, D. L.; and Berrier, B. L.: *Effects of Afterbody Closure and Sting Interference on the Longitudinal Aerodynamic Characteristics of a Fixed-Wing, Twin-Jet Fighter Airplane Model.* NASA TM X-2415, 1971.

10. Maiden, D. L.: *Effect of Airframe Modifications on Longitudinal Aerodynamic Characteristics of a Fixed-Wing, Twin-Jet Fighter Airplane Model.* NASA TM X-2523, 1972.

11. McDonnell Douglas Corp.: *F-15 2-D Nozzle System Integration Study, Volume I—Technical Report.* NASA CR-145295, 1978.

12. McDonnell Douglas Corp.: *F-15 2-D Nozzle System Integration Study, Volume II—2-D Nozzle Effects on the Probability of Detection and Conversion.* NASA CR-145296, 1978.

13. McDonnell Douglas Corp.: *F-15 2-D Nozzle System Integration Study, Volume III–Appendices.* NASA CR-145297, 1978.

14. Pendergraft, O. C.; and Bare, E. Ann: *Fuselage and Nozzle Pressure Distributions on a 1/12-Scale F-15 Propulsion Model at Transonic Speeds.* NASA TP-1521, 1979.

15. Pendergraft, O. C.: *Effect of Nozzle and Vertical-Tail Variables on the Performance of a Three-Surface F-15 at Transonic Mach Numbers.* NASA TP-2043, 1982.

16. Bare, E. Ann; and Pendergraft, Odis C., Jr.: *Effect of Thrust Reverser Operation on the Lateral-Directional Characteristics of a Three-Surface F-15 Model at Transonic Speeds.* NASA TP-2234, 1983.

17. Pendergraft, Odis C.; and Carson, George T.: *Fuselage and Nozzle Pressure Distributions of a 1/12-Scale F-15 Propulsion Model at Transonic Speeds.* NASA TP-2333, 1984.

18. Rivera, Jose A.; and Florance, James R.: Contributions of Transonic Dynamics Tunnel Testing to Airplane Flutter Clearance. AIAA-2300-1768, 2000.

Boeing F-18 Hornet

1. Traven, Ricardo; Hagan, John; and Niewoehner, Robert: Solving Wing Drop on the F/A-18E/F Super Hornet. *Proceedings of the 42nd Symposium of the Society of Experimental Test Pilots—1998 Report to the Aerospace Profession.* Sept. 1998.

2. Capone, Francis J.; Berrier, Bobby L.: *Investigation of Axisymmetric and Nonaxisymmetric Nozzles Installed on a 0.10-Scale F-18 Prototype Airplane Model.* NASA TP-1638, 1980.

3. Bare, E. Ann; Berrier, Bobby L.; and Capone, Francis J.: *Effect of Simulated In-Flight Thrust Reversing on Vertical-Tail Loads of F-18 and F-15 Airplane Models.* NASA TP-1890, 1981.

4. Capone, Francis J.: *Aeropropulsive Characteristics at Mach Numbers Up to 2.2 of Axisymmetric and Nonaxisymmetric Nozzles Installed on an F-18 Model.* NASA TP-2044, 1982.

5. Hwang, C.; Johnson, E.; Mills, G.; Noll, T.; and Farmer, M.: *Wind-Tunnel Test of a Fighter Aircraft Wing/Store Flutter Suppression System, An International Effort.* AGARD-R-689, 1980.

6. Moss, Steven W.; Cole, Stanley R.; and Doggett, Robert V., Jr.: Some Subsonic and Transonic Buffet Characteristics of the Twin-Vertical-Tails of a Fighter Airplane Configuration. AIAA-91-1049, 1991.

7. Moses, Robert W.; and Shah, Gautam H.; Spatial Characteristics of F/A-18 Vertical Tail Buffet Pressures Measured In-Flight. AIAA-98-1956, 1998.

8. Moses, R. W.; and Pendleton, E.: A Comparison of Pressure Measurements Between a Full-Scale and a 1/6-Scale F/A-18 Twin Tail During Buffet. *Loads and Requirements for Military Aircraft,* AGARD-R-815, Feb. 1997, pp. 6-1—6-15.

9. Rivera, Jose A.; and Florance, James R.: Contributions of Transonic Dynamics Tunnel Testing to Airplane Flutter Clearance. AIAA-2000-1768, 2000.

10. Perry Boyd, III; Noll, Thomas E.; and Scott, Robert C.: Contributions of the Transonic Dynamics Tunnel to the Testing of Active Control of Aeroelastic Response. AIAA-2000-1769, 2000.

11. Chambers, Joseph R.; Gilbert, William P.; and Nguyen, Luat T., eds.: *High-Angle-of-Attack Technology—Volume 1.* NASA CP-3149, Vol. 1, Pts. 1,2, & 3, 1992.

12. Chambers, Joseph R.; Burley, James R., II; and Meyer, Robert R., Jr., eds.: *High-Angle-of-Attack Technology—Accomplishments, Lessons Learned, and Future Directions.* NASA/CP-1998-207676, Vol. 1, Pts. 1,2, & 3, 1998.

Boeing T-45 Goshawk

1. Long Development Phase Nearly Over for Goshawk. *Aviat. Week & Space Technol.,* vol. 139, no. 9, Aug. 30, 1993, pp. 38–41.

Fairchild A-10 Thunderbolt II

1. Lockwood, V. E.; and Matarazzo, A.: *Subsonic Wind Tunnel Investigation of a Twin-Engine Attack Airplane Model Having Nonmetric Powered Nacelles.* NASA TN D-7742, 1974.

2. Scher, S. H.; and White, L.: *Spin-Tunnel Investigation of a 1/30-Scale Model of the Fairchild A-10A Airplane.* NASA TM SX-3090, 1974.

General Dynamics F-111 Aardvark

1. Stack, John: Variable-Sweep Aircraft Research in the United States. Memorandum for Director, Foreign Programs (Mutual Weapons Development Program) Feb. 10, 1960.

2. Staff of the Langley Research Center: *Summary of NACA/NASA Variable-Sweep Research and Development Leading to the F-111 (TFX).* NASA LWP-285.

3. Staff of the Langley Research Center: *Summary of NASA Support of the F-111 Development Program. Part I—December 1962–December 1965.* NASA LWP-246, 1966.

4. Polhamus, E. C.: Application of Slender Wing Benefits to Military Aircraft. AIAA-83-2566, 1983.

5. Runckel, Jack F.; Lee, Edwin E., Jr.; and Simonson, Albert J.: *Sting and Jet Interference Effects on the Afterbody Drag of a Twin-Engine Variable-Sweep Fighter Model at Transonic Speeds.* NASA TM X-755, 1963.

6. Kirkham, Frank S.; and Schmeer, James W.: *Performance Characteristics at Mach Numbers up to 1.29 of a Blow-in-Door Ejector Nozzle With Doors Fixed in Full-Open Position.* NASA TM X-830, 1963.

7. Schmeer, James W.; Mercer, Charles E.; and Kirkham, Frank S.: *Effect of Bypass Air on the Performance of a Blow-in-Door Ejector Nozzle at Transonic Speeds.* NASA TM X-896, 1963.

8. Ayers, Theodore G.: *Transonic Aerodynamic Characteristics of a Variable-Wing-Sweep Tactical Fighter Model—Phase 1.* NASA TM X-1039, 1964.

9. Ayers, Theodore G.: *Transonic Aerodynamic Characteristics of a Variable-Wing-Sweep Tactical Fighter Model—Phase 2.* NASA TM X-1040, 1964.

10. Ayers, Theodore G.: *Transonic Aerodynamic Characteristics of a Variable-Wing-Sweep Tactical Fighter Model—Phase 3.* NASA TM X-1183, 1965.

11. Mercer, Charles E.; Schmeer, James W.; and Lauer, Rodney F., Jr.: *Performance of Several Blow-in-Door Ejector Nozzles at Subsonic and Low-Supersonic Speeds.* NASA TM X-1163, 1967.

12. Lauer, Rodney F., Jr.; and Mercer, Charles E.: *Blow-In-Door Ejector Nozzle Performance Comparison Between Fixed-Open and Free-Floating Door Configurations.* NASA TM X-1177, 1967.

13. Re, Richard J.; Wilmoth, Richard G.; and Runckel, Jack F.: *Investigation of Effects of Afterbody Closure and Jet Interference on the Drag of a Twin-Engine Tactical Fighter.* NASA TM X-1382, 1967.

14. Schmeer, James W.; Lauer, Rodney F., Jr.; and Berrier, Bobby L.: *Performance of Blow-in-Door Ejector Nozzles Installed on a Twin-Jet Variable-Wing-Sweep Fighter Airplane Model.* NASA TM X-1383, 1967.

15. Boisseau, Peter C.: *Flight Investigation of Dynamic Stability and Control Characteristics of a 1/10-Scale Model of a Variable-Wing-Sweep Fighter Airplane Configuration.* NASA TM X-1367, 1967.

16. Mercer, Charles E.; Pendergraft, Odis C., Jr.; and Berrier, Bobby L.: *Effect of Geometric Variations on Performance of a Twin-Jet Blow-In-Door Ejector Nozzle Installation.* NASA TM X-1633, 1968.

17. Staff of the Full-Scale Research Division: *NASA LRC Presentation to F-111 Ad Hoc Drag Committee.* NASA LWP-456, 1967.

18. Wilmoth, Richard G.; Norton, Harry T., Jr.; and Corson, Blake W., Jr.: *Effect of Engine-Interfairing Modifications on the Performance of a Powered Twin-Jet Fighter-Airplane Model at Mach 1.20.* NASA TM X-1534, 1968.

19. Norton, Harry T., Jr.: *Installed Performance Characteristics of a Blow-In-Door Ejector Nozzle With Terminal Fairings.* NASA TM X-1611, 1968.

20. Chamales, Nicholas C.; and Salters, Leland B., Jr.: *Effects of Horizontal-Tail Twist and Deflection and of Jet Exhaust on Some Characteristics of a Twin-Jet Fighter Model at Mach 1.20.* NASA TM X-1693, 1968.

21. Bowman, James S., Jr.; and Lee, Henry A.: *Spin-Tunnel Investigation of a 1/40-Scale Model of the F-111A Airplane.* NASA TM SX-1672, U.S. Air Force, 1968.

22. Libbey, Charles E.; and Bowman, James S.: *Radio-Controlled Free-Flight Spin Tests of a 1/9-Scale Model of the F-111A Airplane—COORD No. AF-AM-440.* NASA TM SX-2008, U.S. Air Force, 1970.

23. Chambers, Joseph R.; Anglin, Ernie L.; and Bowman, James S., Jr.: *Effects of a Pointed Nose on Spin Characteristics of a Fighter Airplane Model Including Correlation With Theoretical Calculations.* NASA TN D-5921, 1970.

24. Newman, J. C., Jr.; Elber, W.; and Hudson, C. M.: *Fatigue Crack Propagation in Surface-Cracked D6ac Steel Under Block-Programed Loading.* LWP-930, 1971.

25. Ayers, Theodore G.: *A Wind-Tunnel Investigation of the Application of the NASA Supercritical Airfoil to a Variable-Wing-Sweep Fighter Airplane.* NASA TM X-2759, 1973.

26. Bowman, James S., Jr.; and White, William L.: *Spin-Tunnel Investigation of a 1/40-Scale Model of the F-111A Airplane With Store Loadings and With Supplementary Spin-Recovery Devices—COORD No. AF-AM-440.* NASA TM SX-2970, U.S. Air Force, 1974.

27. Coulam, Robert F.; and McNamara, Robert S.: *Illusions of Choice—The F-111 and the Problem of Weapons Acquisition Reform.* Princeton Univ. Press, 1977.

28. Hallissy, James B.; and Ayers, Theodore G.: *Transonic Wind-Tunnel Investigation of the Maneuver Potential of the NASA Supercritical Wing Concept—Phase I.* NASA TM X-3534, 1977.

29. Ayers, Theodore G.; and Hallissy, James B.: *Historical Background and Design Evolution of the Transonic Aircraft Technology Supercritical Wing.* NASA TM-81356, 1981.

30. Bonnema, Kenneth L.; and Smith, Stephen B.: *AFTI/F-111 Mission Adaptive Wing Flight Research Program.* AIAA-88-2118, 1988.

31. Committee on Aging of U.S. Air Force Aircraft: *Aging of U.S. Air Force Aircraft—Final Report.* Natl. Acad. Press, 1997.

32. Hallion, Richard P.: *On the Frontier—Flight Research at Dryden, 1946–1981.* NASA SP-4303, 1984.

33. Rivera, Jose A.; and Florance, James R.: Contributions of Transonic Dynamics Tunnel Testing to Airplane Flutter Clearance. AIAA-2000-1768, 2000.

Grumman A-6 Intruder

1. Mercer, Charles E.; Salters, Leland B., Jr.; and Capone, Francis J.: *Afterbody Temperatures, Pressures, and Aerodynamic Characteristics Resulting From Extension of Speed-Brake Configurations Into the Exhaust Jets of a Twin-Engine Attack-Type-Airplane Model.* NASA TM X-517, 1961.

2. Lee, Edwin E., Jr.; and Mercer, Charles E.: *Jet Interference Effects on a Twin-Engine Attack-Type-Airplane Model With Large Speed-Brake, Thrust-Spoiler Surfaces.* NASA TM X-454, 1961.

3. Lee, Henry A.; and Healy, Frederick M.: *Spin-Tunnel Investigation of a 1/28-Scale Model of a Subsonic Attack Airplane—COORD No. N-AM-67.* NASA TM SX-964, Bureau of Weapons, Dept. of Navy, 1964.

4. Cole, Stanley R.; Rivera, Jose A., Jr.; and Nagaraja, K. S.: Flutter Study of an Advanced Composite Wing With External Stores. AIAA-87-0880, 1987.

5. Klemens, Susan M.: *The A-6: Winning Ugly.* http://www.discovery.com/area/technology/a6/intruder.html Accessed Mar. 23, 2000.

6. Rivera, Jose A.; and Florance, James R.: Contributions of Transonic Dynamics Tunnel Testing to Airplane Flutter Clearance. AIAA-2000-1768, 2000.

Grumman EA-6B Prowler

1. Lee, Henry A.: *Spin-Tunnel Investigation of a 1/28-Scale Model of the Grumman EA-6B Airplane—COORD No. N-AM-145.* NASA TM SX-2249, U.S. Navy, 1971.

2. Jordan, Frank L., Jr.; Hahne, David E.; Masiello, Matthew F.; and Gato, William: High-Angle-of-Attack Stability and Control Improvements for the EA-6B Prowler. *A Collection of Technical Papers—AIAA 5th Applied Aerodynamics Conference,* Aug. 1987, pp. 270–285. (Available as AIAA-87-2361.)

3. Sewall, W. G.; McGhee, R. J.; and Ferris, J. C.: Wind-Tunnel Test Results of Airfoil Modifications for the EA-6B. *A Collection of Technical Papers—AIAA 5th Applied Aerodynamics Conference,* Aug. 1987, pp. 248–256. (Available as AIAA-87-2359.)

4. Waggoner, E. G.; and Allison, D. O.: EA-6B High-Lift Wing Modifications. *A Collection of Technical Papers—AIAA 5th Applied Aerodynamics Conference,* Aug. 1987, pp. 257–269. (Available as AIAA-87-2360.)

5. Gato, W.; and Masiello, M. F.: Innovative Aerodynamics—The Sensible Way of Restoring Growth Capability to the EA-6B Prowler. AIAA-87-2362, 1987.

6. Hanley, Robert J.: Development of an Airframe Modification To Improve the Mission Effectiveness of the EA-6B Airplane. *A Collection of Technical Papers—AIAA 5th Applied Aerodynamics Conference,* Aug. 1987, pp. 241–247. (Available as AIAA-87-2358.)

Grumman F-14 Tomcat

1. Harris, Roy V.; et. al.: Summary of Langley Contributions to the F-14, F-15, F-16, and F-18 Fighter Aircraft. Memorandum to the Director from Chief, High-Speed Aerodynamics Division. NASA Internal Memo, Aug. 26, 1981.

2. Staff of the Full-Scale Research Division: *The LFAX-4A Fighter/Attack Airplane Concept.* NASA LWP-518, 1967.

3. Alford, William J.; and Henderson, William P.: *NASA Assessment of the United States Navy F-14 Airplane Designs.* NASA LWP-794, 1969.

4. Assistant Secretary of the Navy: Letter to NASA Administrator. May 1968.

5. Putnam, Lawrence E.: *Effect of Nozzle Interfairings on Aerodynamic Characteristics of a Twin-Engine Variable-Sweep-Wing Fighter Airplane at Mach 0.60 to 2.01.* NASA TM X-2769, 1973.

6. Mercer, Charles E.; and Reubush, David E.: *Sting and Jet Interference Effects on Longitudinal Aerodynamic Characteristics of a Twin-Jet, Variable-Wing-Sweep Fighter Model at Mach Numbers to 2.2.* NASA TM X-2825, 1973.

7. Putnam, Lawrence E.: *Effects of Modifications and External Stores on Aerodynamic Characteristics of a Fighter Airplane With Variable-Sweep Wing and Twin Vertical Tails at Mach 0.65 to 2.01.* NASA TM X-2366, 1974.

8. Reubush, David E.; and Mercer, Charles E.: *Exhaust-Nozzle Characteristics for a Twin-Jet Variable-Wing-Sweep Fighter Airplane Model at Mach Numbers to 2.2.* NASA TM X-2947, 1974.

9. Reubush, David E.; and Mercer, Charles E.: *Effects of Nozzle Interfairing Modifications on Longitudinal Aerodynamic Characteristics of a Twin-Jet, Variable-Wing-Sweep Fighter Model.* NASA TN D-7817, 1975.

10. Reubush, David E.; and Carlson, John R.: *Effects of Installation of F101 DFE Exhaust Nozzles on the Afterbody-Nozzle Characteristics of the F-14 Airplane.* NASA TM-83250, 1982.

11. Reubush, David E.; and Berrier, Bobby L.: *Effects of the Installation and Operation of Jet-Exhaust Yaw Vanes on the Longitudinal and Lateral-Directional Characteristics of the F-14 Airplane.* NASA TP-2769, 1987.

12. Rivera, Jose A.; and Florance, James R.: Contributions of Transonic Dynamics Tunnel Testing to Airplane Flutter Clearance. AIAA-2000-1768, 2000.

Grumman X-29 Advanced Technology Demonstrator

1. Diederich, Franklin W.; and Budiansky, Bernard: *Divergence of Swept Wings.* NACA TN 1680, 1948.

2. Harris, Thomas A.: *Documentation of the 300-Mile Per Hour 7- by 10-Foot Wind Tunnel.* NASA Internal Document, NASA Langley Research Center, Mar. 1972.

3. Krone, Norris J., Jr.: Divergence Elimination With Advance Composites. AIAA-75-1009, Aug. 1975.

4. Ricketts, Rodney H.; and Doggett, Robert V., Jr.: *Wind-Tunnel Experiments on Divergence of Forward-Swept Wings.* NASA TP-1685, 1980.

5. Doggett, Robert V., Jr.; and Ricketts, Rodney H.: *Dynamic Response of a Forward-Swept-Wing Model at Angles of Attack up to 15° at a Mach Number of 0.8.* NASA TM-81863, 1980.

6. Murri, D. G.; Croom, M. A.; and Nguyen, L. T.: High Angle-of-Attack Flight Dynamics of a Forward-Swept Wing Fighter Configuration. AIAA-83-1837, 1983.

7. Murri, Daniel G.; Nguyen, Luat T.; and Grafton, Sue B.: *Wind-Tunnel Free-Flight Investigation of a Model of a Forward-Swept-Wing Fighter Configuration.* NASA TP-2230, 1984.

8. Chipman, R.; Rauch, F.; Rimer, M.; Muniz, B.; and Ricketts, R. H.: Transonic Test of a Forward Swept Wing Configuration Exhibiting Body Freedom Flutter. AIAA-85-0689, 1985.

9. Fratello, David J.; Croom, Mark A.; Nguyen, Luat T.; and Domack, Christopher S.: Use of the Updated NASA Langley Radio-Controlled Drop-Model Technique for High-Alpha Studies of the X-29A Configuration. *A Collection of Technical Papers—AIAA Atmospheric Flight Mechanics Conference,* Aug. 1987, pp. 305–317. (Available as AIAA-87-2559.)

10. Raney, David L.; and Batterson, James G.: *Lateral Stability Analysis for X-29A Drop Model Using System Identification Methodology.* NASA TM-4108, 1989.

11. Rivera, Jose A.; and Florance, James R.: Contributions of Transonic Dynamics Tunnel Testing to Airplane Flutter Clearance. AIAA-2000-1768, 2000.

Lockheed Martin C-130 Hercules

1. Petit, P. H.: *An Applications Study of Advanced Composite Materials to the C-130 Center Wing Box.* Lockheed-Georgia Co., NASA CR-66979, 1970.

2. Yost, J. D.: Correction Factors for Miner's Fatigue Damage Equation Derived From C-130 Fleet Aircraft Fatigue Cracks. AIAA-86-2684, 1986.

3. NASA and Industry-Partners in C-130 Technology. *Researcher News, vol. 12, issue 23.* Langley Research Center, Nov. 27, 1998.

Lockheed Martin C-141 Starlifter

1. Ruhlin, Charles L.; Sandford, Maynard C.; and Yates, E. Carson, Jr.: *Wind-Tunnel Flutter Studies of the Sweptback T-Tail of a Large Multijet Cargo Airplane at Mach Numbers to 9.90.* NASA TN D-2179, 1964.

2. Sandford, Maynard C.; Ruhlin, Charles L.; and Yates, E. Carson, Jr.: *Subsonic and Transonic Flutter and Flow Investigations of the T-Tail of a Large Multijet Cargo Airplane.* NASA TN D-4316, 1968.

3. Sandford, Maynard C.; and Ruhlin, Charles L.: *Wind-Tunnel Study of Deflected-Elevator Flutter Encountered on a T-Tail Airplane.* NASA TN D-5024, 1969.

4. Abel, Irving: *Evaluation of a Technique for Determining Airplane Aileron Effectiveness and Roll Rate by Using an Aeroelastically Scaled Model.* NASA TN D-5538, 1969.

5. Cleveland, F. A.; and Gilson, R. D.: Development Highlights of the C-141 Starlifter. AIAA-64-596, 1965.

6. Victor Crash. *Aviat. Week & Space Technol.,* vol. 61, no. 4, July 26, 1954, p. 16.

7. Loving, Donald L.: *Wind-Tunnel—Flight Correlation of Shock-Induced Separated Flow.* NASA TN D-3580, 1966.

8. MacWilkinson, D. G.; Blackerby, W. T.; and Paterson, J. H.: *Correlation of Full-Scale Drag Predictions With Flight Measurements on the C-141A Aircraft—Phase II, Wind Tunnel Test, Analysis, and Prediction Techniques. Volume 1—Drag Predictions, Wind Tunnel Data Analysis and Correlation.* NASA CR-2333, 1974.

9. Rivera, Jose A.; and Florance, James R.: Contributions of Transonic Dynamics Tunnel Testing to Airplane Flutter Clearance. AIAA-2000-1768, 2000.

Lockheed Martin C-5 Galaxy

1. Luoma, Arvo A.; Re, Richard J.; and Loving, Donald L.: *Subsonic Longitudinal Aerodynamic Measurements on a Transport Model in Two Slotted Tunnels Differing in Size.* NASA TM X-1660, 1968.

2. Loving Donald L.; and Luoma, Arvo A.: *Sting-Support Interference on Longitudinal Aerodynamic Characteristics of Cargo-Type Airplane Models at Mach 0.70 to 0.84.* NASA TN D-4021, 1967.

3. Patterson, James C., Jr.; and Flechner, Stuart G.: *Jet-Wake Effect of a High-Bypass Engine on Wing-Nacelle Interference Drag of a Subsonic Transport Airplane.* NASA TN D-6067, 1970.

4. Parlett, Lysle P.; Fink, Marvin P.; and Freeman, Delma C., Jr. (appendix B by Marion O. McKinney and Joseph L. Johnson, Jr.): *Wind-Tunnel Investigation of Large Jet Transport Model Equipped With an External-Flow Jet Flap.* NASA TN D-4928, 1968.

5. Grantham, Willaim D.; Deal, Perry L.; and Sommer, Robert W.: *Simulator Study of the Instrument Landing Approach of a Heavy Subsonic Jet Transport With an External-Flow Jet-Flap System Used for Additional Lift.* NASA TN D-5862, 1970.

6. Ruhlin, Charles L.; and Sandford, Maynard C.: *Experimental Parametric Studies of Transonic T-Tail Flutter.* NASA TN D-8066, 1975.

7. Thompson, William C.: *Ditching Investigation of a 1/30-Scale Dynamic Model of a Heavy Jet Transport Airplane.* NASA TM X-2445, 1972.

8. Blackwell, James A., Jr.: *Preliminary Study of Effects of Reynolds Number and Boundary-Layer Transition Location on Shock-Induced Separation.* NASA TN D-5003, 1969.

9. McWhirter, H. D.; Hollenback, W. W.; and Grosser, W. F.: *Correlation of C-5A Active Lift Distribution Control System (ALDCS) Aeroelastic Model and Airplane Flight Test Results.* NASA CR-144903, 1976.

10. Rivera, Jose A.; and Florance, James R.: Contributions of Transonic Dynamics Tunnel Testing to Airplane Flutter Clearance. AIAA-2000-1768, 2000.

11. Perry, Boyd, III; Noll, Thomas E.; and Scott, Robert C.: Contributions of the Transonic Dynamics Tunnel to the Testing of Active Control of Aeroelastic Response. AIAA-2000-1769, 2000.

Lockheed Martin F-16 Fighting Falcon

1. Jackson, Charles M.: NASA/General Dynamics Cooperative Research Leading to and Involving SCAMP. NASA Internal Report, May 1980.

2. Polhamus, E. C.: Application of Slender Wing Benefits to Military Aircraft. AIAA-83-2566, 1983.

3. Buckner, J. K.; Benepe, D. B.; and Hill, P. W.: Aerodynamic Design Evolution of the YF-16. AIAA-74-935, Aug. 1974.

4. Nguyen, Luat T.; Ogburn, Marilyn E.; Gilbert, William P.; Kibler, Kemper S.; Brown, Phillip W.; and Deal, Perry L.: *Stimulator Study of Stall/Post-Stall Characteristics of a Fighter Airplane With Relaxed Longitudinal Static Stability.* NASA TP-1538, 1979.

5. Peloubet, R. P., Jr.; and Haller, R. L.: Wind-Tunnel Demonstration of Active Flutter Suppression Using F-16 Model With Stores. AFWAL-TR 83-3046, Vol.1, 1983.

6. Foughner, J. T., Jr.; and Bensinger, C. T.: F-16 Flutter Model Studies With External Wing Stores. NASA TM-74078, 1977.

7. Perry, Boyd, III; Noll, Thomas E.; and Scott, Robert C.: Contributions of the Transonic Dynamics Tunnel to the Testing of Active Control of Aeroelastic Response. AIAA-2000-1769, 2000.

Lockheed Martin F-22 Raptor

1. Mullin, Sherman N.: The Evolution of the F-22 Advanced Tactical Fighter. Wright Brothers Lecture, AIAA-92-4188, 1992.

2. Aronstein, David C.; Hirchberg, Michael J.; and Piccirillo, Albert C.: Advanced Tactical Fighter to F-22 Raptor—Origins of the 21st Century Air Dominance Fighter. AIAA, 1998.

Lockheed Martin P-3 Orion

1. Baals, Donald D.; and Corliss, William R.: *Wind Tunnels of NASA*. NASA SP-440, 1981.

2. Abbot, F. T., Jr; Kelly, H. Neale; and Hampton, Kenneth D.: *Investigation of Propeller-Power-Plant Auto-Precession Boundaries for a Dynamic-Aeroelastic Model of a Four-Engine Turboprop Transport Airplane*. NASA TN D-1806, June 1963.

3. Rivera, Jose A.; and Florance, James R.: Contributions of Transonic Dynamics Tunnel Testing to Airplane Flutter Clearance. AIAA-2000-1768, 2000.

Lockheed Martin S-3 Viking

1. Lee, Henry A.; and White, William L.: *Spin-Tunnel Investigation of a 1/32-Scale Model of the Lockheed S-3A Airplane—COORD No. N-AM-161*. NASA TM SX-3057, U.S. Navy, 1974.

2. Rivera, Jose A.; and Florance, James R.: Contributions of Transonic Dynamics Tunnel Testing to Airplane Flutter Clearance. AIAA-2000-1768, 2000.

McDonnell Douglas F-4 Phantom II

1. Carmel, Melvin M.; and Gregory, Donald T.: *Preliminary Investigation of the Static Longitudinal and Lateral Stability Characteristics of a Model of a 45° Swept Wing Airplane at Mach Numbers of 1.59, 1.89, and 2.09*. NASA MEMO 3-30-59L, 1959.

2. Carmel, Melvin M.; and Turner, Kenneth L.: *Investigation of Drag and Static Longitudinal and Lateral Stability and Control Characteristics of a Model of a 45° Swept Wing Airplane at Mach Numbers of 1.57, 1.87, 2.16, and 2.53, Phase II Model*. NASA MEMO 3-31-59L, 1959.

3. Oehman, Waldo I.; and Turner, Kenneth L.: *Aerodynamic Characteristics of a 45° Swept-Wing Fighter-Airplane Model and Aerodynamic Loads on Adjacent Stores and Missiles at Mach Numbers of 1.57, 1.87, 2.16, and 2.53*. NACA RM L58C17, 1958.

4. Chambers, Joseph R.; and Anglin, Ernie L.: *Analysis of Lateral-Directional Stability Characteristics of a Twin-Jet Fighter Airplane at High Angles of Attack*. NASA TN D-5361, 1969.

5. Chambers, Joseph R.; Bowman, James S., Jr.; and Anglin, Ernie L.: *Analysis of the Flat-Spin Characteristics of a Twin-Jet Swept-Wing Fighter Airplane*. NASA TN D-5409, 1969.

6. Ray, Edward J.; and Hollingsworth, Eddie G.: *Subsonic Characteristics of a Twin-Jet Swept-Wing Fighter Model With Maneuvering Devices*. NASA TN D-6921, 1973.

7. Moore, Frederick L.; Anglin, Ernie L.; Adams, Mary S.; Deal, Perry L.; and Person, Lee H., Jr.: *Utilization of a Fixed-Base Simulator To Study the Stall and Spin Characteristics of Fighter Airplanes*. NASA TN D-6117, 1971.

8. Newsom, William A., Jr.; and Grafton, Sue B.: *Free-Flight Investigation of Effects of Slats on Lateral-Directional Stability of a 0.13-Scale Model of the F-4E Airplane—COORD No. AF-AM-113*. NASA TM SX-2337, U.S. Air Force, 1971.

9. Doggett, R. V.; and Hanson, P. H.: *Wind Tunnel Buffet Pressure Investigation on the Lower Nose Portion of the RF-4C Aircraft*. NASA LWP-227, 1966.

Missiles

1. Blair, A. B., Jr.; Allen, Jerry M.; and Hernandez, Gloria: *Effect of Tail-Fin Span on Stability and Control Characteristics of a Canard-Controlled Missile at Supersonic Mach Numbers*. NASA TP-2157, 1983.

2. Blair, A. B., Jr.; and Rapp, G. H.: Experimental and Analytical Comparison of Aerodynamic Characteristics of a Forward-Control Missile. AIAA-80-0374, 1980.

Rockwell B-1 Lancer

1. Newsom, William A., Jr.; and Grafton, Sue B.: *Free-Flight Investigation of a 1/17-Scale Model of the B-1 Airplane at High Angles of Attack—COORD No. AF-AM-128.* NASA TM SX-2744, U.S. Air Force, 1973.

2. Grantham, William D.; Deal, Perry L.; and Libbey, Charles E.: *Piloted Simulator Study of the Stability and Control Characteristics of the B-1 Airplane at High Angles of Attack—COORD No. AF-AM-128.* NASA TM SX-3381, U.S. Air Force, 1976.

3. Re, Richard J.; and Reubush, David E.: *Effect of Several Airframe/Nozzle Modifications on the Drag of a Variable-Sweep Bomber Configuration.* NASA TM-80129, 1979.

4. Seiner, John M.; Manning, James C.; Capone, Francis J.; and Pendergraft, Odis C., Jr.: *Study of External Dynamic Flap Loads on a 6 Percent B-1B Model.* ASME-PAPER-91-GT-236, 1991.

5. Rivera, Jose A.; and Florance, James R.: Contributions of Transonic Dynamics Tunnel Testing to Airplane Flutter Clearance. AIAA-2000-1768, 2000.

Rockwell X-31 Enhanced Fighter Maneuverability Demonstrator

1. Croom, M. A.; and Schnellenger, H. G.: *High-Alpha Flight Dynamics of the X-31 Configuration.* NASA CP-3149, 1990.

2. Knox, Fred D.: X-31 Flight Test Update. AIAA-92-1035, 1992.

3. Banks, Daniel W.; Gatlin, Gregory M.; and Paulson, John W., Jr.: *Low-Speed Longitudinal and Lateral-Directional Aerodynamic Characteristics of the X-31 Configuration.* NASA TM-4351, 1992.

4. Scott, William B.: X-31 Completes Post-Stall Test. *Aviat. Week & Space Technol.*, vol. 138, no. 20, May 17 1993, pp. 29–30.

5. Croom, Mark A.; Fratello, David J.; Whipple, Raymond D.; O'Rourke, Matthew J.; and Trilling, Todd W.: Dynamic Model Testing of the X-31 Configuration for High-Angle-of-Attack Flight Dynamics Research. AIAA-93-3674, 1993.

6. Cobleigh, Brent R.: *High-Angle-of-Attack Yawing Moment Asymmetry of the X-31 Aircraft From Flight Test.* NASA CR-186030, 1994.

Appendix

1. Perry, Boyd, III; Noll, Thomas E.; and Scott, Robert C.: Contributions of the Transonic Dynamics Tunnel to the Testing of Active Control of Aeroelastic Response. AIAA-2000-1769, 2000.

2. Cole, Stanley R.; and Garcia, Jerry L.: Past, Present, and Future Capabilities of the Transonic Dynamics Tunnel from an Aeroelastic Perspective. AIAA-2000-1767, 2000.

3. Schuster, David M.; Edwards, John W.; and Bennett, Robert M.: An Overview of Unsteady Pressure Measurements in the Transonic Dynamics Tunnel. AIAA-2000-1770, 2000.

4. Rivera, Jose A.; and Florance, James R.: Contributions of Transonic Dynamics Tunnel Testing to Airplane Flutter Clearance. AIAA-2000-1768, 2000.

5. Baals, D. D.; and Corliss, W. R.: *Wind Tunnels of NASA.* NASA SP-440, 1981.

Index

F

Monographs in Aerospace History

Launius, Roger D., and Gillette, Aaron K. Compilers. *The Space Shuttle: An Annotated Bibliography.* (Monographs in Aerospace History, No. 1, 1992).

Launius, Roger D., and Hunley, J.D. Compilers. *An Annotated Bibliography of the Apollo Program.* (Monographs in Aerospace History, No. 2, 1994).

Launius, Roger D. *Apollo: A Retrospective Analysis.* (Monographs in Aerospace History, No. 3, 1994).

Hansen, James R. *Enchanted Rendezvous: John C. Houbolt and the Genesis of the Lunar-Orbit Rendezvous Concept.* (Monographs in Aerospace History, No. 4, 1995).

Gorn, Michael H. *Hugh L. Dryden's Career in Aviation and Space.* (Monographs in Aerospace History, No. 5, 1996).

Powers, Sheryll Goecke. *Women in Aeronautical Engineering at the Dryden Flight Research Center, 1946-1994* (Monographs in Aerospace History, No. 6, 1997).

Portree, David S.F. and Trevino, Robert C. *Compilers. Walking to Olympus: A Chronology of Extravehicular Activity (EVA).* (Monographs in Aerospace History, No. 7, 1997).

Logsdon, John M. *Moderator. The Legislative Origins of the National Aeronautics and Space Act of 1958: Proceedings of an Oral History Workshop* (Monographs in Aerospace History, No. 8, 1998).

Rumerman, Judy A. *Compiler. U.S. Human Spaceflight: A Record of Achievement, 1961-1998* (Monographs in Aerospace History, No. 9, 1998).

Portree, David S.F. *NASA's Origins and the Dawn of the Space Age* (Monographs in Aerospace History, No. 10, 1998).

Logsdon, John M. *Together in Orbit: The Origins of International Cooperation in the Space Station Program* (Monographs in Aerospace History, No. 11, 1998).

Phillips, W. Hewitt. *Journey in Aeronautical Research: A Career at NASA Langley Research Center* (Monographs in Aerospace History, No. 12, 1998).

Braslow, Albert L. *A History of Suction-Type Laminar-Flow Control with Emphasis on Flight Research* (Monographs in Aerospace History, No. 13, 1999).

Logsdon, John M. *Moderator. Managing the Moon Program: Lessons Learned from Project Apollo* (Monographs in Aerospace History, No. 14, 1999).

Perminov, V.G. *The Difficult Road to Mars: A Brief History of Mars Exploration in the Soviet Union* (Monographs in Aerospace History, No. 15, 1999).

Tucker, Tom. *Touchdown: The Development of Propulsion Controlled Aircraft at NASA Dryden* (Monographs in Aerospace History, No. 16, 1999).

Maisel, Martin D.; Demo J. Giulianetti; and Daniel C. Dugan. *The History of the XV-15 Tilt Rotor Research Aircraft: From Concept to Flight.* (Monographs in Aerospace History, NASA SP-2000-4517, 2000).

Jenkins, Dennis R. *Hypersonics Before the Shuttle: A History of the X-15 Research Airplane.* (Monographs in Aerospace History, NASA SP-2000-4518, 2000).

About the Author

Joseph R. Chambers is an aviation consultant who lives in Yorktown, Virginia. He retired from the NASA Langley Research Center in 1998 after a 36-year career as a researcher and manager of military and civil aeronautics research activities. He began his career as a specialist in flight dynamics as a member of the staff of the Langley 30-by 60-Foot (Full-Scale) Tunnel, where he conducted research on a variety of aerospace vehicles including V/STOL configurations, reentry vehicles, and fighter aircraft configurations. He later became a division chief with responsibilities for all research projects in the Full-Scale Tunnel, the 20-Foot Vertical Spin Tunnel, drop-model tests, and piloted simulator studies in the Differential Maneuvering Simulator (DMS) at Langley. When he retired from NASA, he was manager of a division that conducted systems analysis of the potential payoffs of advanced aircraft concepts and research investments.

Mr. Chambers participated in the development of virtually every Air Force and Navy aircraft of the past 30 years, including all fighter-attack aircraft. He is the author of over 50 technical reports and publications and coauthor of NASA Special Publication SP-514 on the subject of airflow condensation patterns for high-performance aircraft. He has made presentations on research and aircraft development programs to audiences as diverse as the Von Karman Institute in Belgium and the annual Experimental Aircraft Association (EAA) Fly-In at Oshkosh, WI. He has served as a representative of the United States on international committees for military research and has given presentations in Japan, China, Australia, the United Kingdom, Canada, Italy, France, Germany, and Sweden.

Mr. Chambers received the Exceptional Service Medal and the Outstanding Leadership Medal, two of NASA's highest awards. He received the Arthur Flemming Award in 1975 as one of the 10 Most Outstanding Civil Servants for his management of NASA stall-spin research for military and civil aircraft. He also was a corecipient of the 1973 H.J.E. Reid Award presented by Langley for the most outstanding technical paper (an analysis of the lateral-directional stability of the F-4 fighter aircraft at high angles of attack). In 1977, he was a corecipient of the Mechanics and Control of Flight Award of the American Institute of Aeronautics and Astronautics. This awarded was presented for his contributions to technology to predict high-angle-of-attack, stall, and spin characteristics for aircraft and for improving those characteristics through combined aerodynamic and automatic control design. Mr. Chambers has a bachelor of science degree in Aeronautical Engineering from the Georgia Institute of Technology and a master of science degree in Aerospace Engineering from Virginia Polytechnic Institute and State University (Virginia Tech).

www.ingramcontent.com/pod-product-compliance
Lightning Source LLC
Chambersburg PA
CBHW081436170526
45166CB00008B/2216